T0341303

HEALTH FOODS FROM OCEAN ANIMALS

Marine animals and their body constituents have been in use by mankind for nutrition and medical applications centuries ago. This book contains some well known and lesser known compounds from some important marine animals those have been consumed by man for centuries. This is the first book in this field and will serve as a reference for future researchers in the field.

Prof. K. Gopakumar, Professor of Eminence and Head of Department, Food Science and Technology is a Ph.D in Biochemistry. He has served as the Deputy Director General (Fisheries) of the Indian Council for Agriculture Research (ICAR) and Director, Central Institute of Fisheries Technology (CIFT), Kochi. He served as Chief Technical Advisor to the Philippines Government for Food and Agriculture Organization (FAO) of the United Nations (UN) and also expert of FAO in Bangladesh and UN Mission member to Sri Lanka. He did his Post Doctoral Research in Natural Resources Institute, London and did advanced training in Retort Pouch Processing at Metal Box, UK.

Dr. Balgopal Gopakumar, son of Dr Gopakumar, is a Scientist at Memorial Sloan Kettering Cancer Center in Manhattan, New York. His research has been primarily based on the fields of data analytics, process and quality in the field of healthcare. Dr. Balagopal has over 10 years of research experience in process & quality improvement, analytics, Lean, Six-Sigma in the field of healthcare systems.

HEALTH FOODS FROM OCEAN ANIMALS

By

K. Gopakumar

Ph.D, D.Sc, FIC, FNAAS, FSFT, FAA
Former Deputy Director General
Indian Council of Agricultural Research, New Delhi and,
Director, Central Institute of Fisheries Technology, Cochin and
Currently Professor of Eminence, Department of Food science and Technology
Kerala University of Fisheries and Ocean studies,
COCHIN-682506

and

Balagopal Gopakumar

MS, PhD (SUNY, USA)
Scientist
Memorial Sloan Kettering New York,
USA

CRC Press
Taylor & Francis Group
Boca Raton London New York

CRC Press is an imprint of the
Taylor & Francis Group, an **Informa** business

**NARENDRA PUBLISHING
HOUSE**
DELHI (INDIA)

First published 2021
by CRC Press
2 Park Square, Milton Park, Abingdon, Oxon, OX14 4RN
and by CRC Press
6000 Broken Sound Parkway NW, Suite 300, Boca Raton, FL 33487-2742

© 2021 Narendra Publishing House

CRC Press is an imprint of Informa UK Limited

The right of K. Gopakumar and Balagopal Gopakumar to be identified as authors of this work has been asserted by them in accordance with sections 77 and 78 of the Copyright, Designs and Patents Act 1988.

Print edition not for sale in South Asia (India, Sri Lanka, Nepal, Bangladesh, Pakistan or Bhutan).

British Library Cataloguing-in-Publication Data
A catalogue record for this book is available from the British Library

Library of Congress Cataloging-in-Publication Data
A catalog record has been requested

ISBN: 978-0-367-54049-4 (hbk)
ISBN: 978-1-003-08424-2 (ebk)

NARENDRA PUBLISHING
HOUSE
DELHI (INDIA)

Contents

Preface

It is very difficult to precisely establish the exact period in which man started using marine animals and their body parts for medical applications. The oldest literature describing the medical applications of marine animals is the Chinese Medical Dictionary called *Chinese Materia Medica* (CMM). Its origin dates back to 1100 BCE. The Yellow Emperor's Cannon (*Huang Di Nei Jing)* is another *Chinese medical Theory and* it was estimated compiled in the first century BCE.

The traditional Indian Medical treatise called Ayurveda includes *Sushruta Samhita* and *Charaka Samhita.* Both were compiled around early parts of AD. Both the traditional Chinese and Indian systems of medicine use plant and their roots, leaves and fruits as well as their extracts for treating various ailments that afflict both man and animals. The Chinese traditional medical practioners used marine animals and their bones (eg. shark bones) and sea weeds for therapeutical applications. The Ayurveda physicians used various animal products like milk, bones and gallstones for medical use. There are also evidences showing that they used fish and fish oils like shark liver oil for preparation drugs.

There is nothing new in this book. Certain work in the past and present are collected and collated for the young scientists to understand wonderful uses of marine animals and to undertake more investigations in this field. It is written in semi-scientific language so that anybody can read it. A large number of references are added at the end of each chapter so that research scientists can use them to expand their knowledge. We hope that it will be well received by the public and scientists in this field.

K. Gopakumar

G. Balagopal

Central Agricultural University
Lamphel Pat, Imphal-795004
Manipur

Prof. S. Ayyappan
Chancellor

Foreword

Marine organisms and their body parts like bone and liver oil of some fishes have been in use by man for centuries for nutraceutical and medical applications. A large number marine organisms like oysters, clams, mussels and fish oils like cod and shark liver oils have been in use by man even before the discovery of biomolecules present in them. The science behind them was not known to us until the dawn of the 20th century when organic chemists started identifying the active chemicals in them. These developments propelled many scientists to undertake investigations in this field. This has resulted in the isolation and synthesis of several compounds which facilitated clinical research in this field.

In the past, traditional medical physicians used marine animals for the treatment of ailments based on the knowledge passed on to them by their predecessors based on traditional wisdom and possibly supported by results from patients who used them. During the latter half of the 20th century, extensive clinical trials were conducted by wellknown research centres and found that most of them failed to get clinical supports from Agencies like USFDA, WHO or other international agencies. However, even without support from clinical studies, thousands of marine nutraceuticals are sold all over the world and they enjoy huge markets. Obviously, the users would have got some benefits or cure for many dreaded diseases like cancer and AIDS, for there are no proven drugs even today.Hence, there is a need for more intensive research in this field.

Mobile: 0091-9582898989; e-mail: sayyappan1995@gmail.com
Residence: 172, Shreepadam, 5th Main Road, Avalahalli,
BDA Extension, Bengaluru – 560085, Karnataka

Central Agricultural University
Lamphel Pat, Imphal-795004
Manipur

In this context, this book first of its kind, is an excellent compilation of nutritional and medical applications of marine animals undertaken in the past and present. The authors have collected a large number data on composition, nutritional and medical use of marine animals. One of the authors, Dr K.Gopakumar is the most reputed authority in the country in this field. It is he who initiated investigations in this field in India more than fifty years ago at the Central Institute of Fisheries Technology,Kochi. He has authored more than 350 research papers in international and national journals of high impact. He is also the winner of the nation's top scientific award of the Indian Council of Agricultural Research(ICAR), the Rafi AhmedKidwai Award and the most reputed international award, the Kwarizimi International Award given by the Islamic Republic of Iran in 2001, for his research work on the chemistry of natural products. The book is coauthored by his son, Dr Balagopal. I am sure this will be an excellent reference book for many young scientists who desire to undertake research in the country and in the world.

25 August 2018

(S. Ayyappan)

Mobile: 0091-9582898989; e-mail: sayyappan1995@gmail.com
Residence: 172, Shreepadam, 5th Main Road, Avalahalli,
BDA Extension, Bengaluru – 560085, Karnataka

Acknowledgements

We are grateful Dr. Ravisankar, Director, Central Institute of Fisheries Technology (CIFT), Kochi for granting permission to utilize some of the published data on composition of fish and crustaceans. Dr. Susheela Mathew, Head Department of Biochemistry and Nutrition Division for undertaking analysis of oyster meat and supplying data on biochemical composition of oysters. We are also indebted to Professor Venkitaramani, Kerala University of Fisheries and Ocean Sciences (KUFOS) and former Dean Faculty of Fish Fisheries College, Tuticorin, India , an authority on systematics, for verifying the technical names of crustaceans mentioned in the book and to Dr. Suresh Kumar, Professor and Director, School of Ocean Studies and Technology (KUFOS) for supplying live photographs of sea cucumbers and octopus.

We are pleased to express our gratitude to Ms. Mary Ann Wagner, Assistant Director for Communications, Washington Sea Grant College of Engineering, University of Washington, U.S.A. for granting permission to use compositional data of crustaceans.

We are thankful to Professor Sreedharan, Managing Director, Aasha Biochem. Pvt. Ltd., Vatakara, Kerala- 673 106, India for supplying technical data on composition of sqalene and alkoxy glycerol and permission to publish the data.

Mr. Nikhil Rai , Managing Director, Perma Healthcare, New Delhi -110044 has supplied valuable information on seatone. This help is gratefully acknowledged.

We are also grateful to Dr. S. Ayyappan, former Director General, Indian Council of Agricultural Research and Secretary, Department of Agriculture and Education, Ministry of Agriculture, Government of India for encouraging us to write this book and for writing a fitting forward to this book. The help given by Professor P.T. Mathew and encouragement and support given by Dr. A. Ramachandran, Vice-Chancellor KUFOS are gratefully acknowledged.

K. Gopakumar

G. Balagopal

List of Tables

List of Figures

CHAPTER **1**

SQUALENE

1.1 INTRODUCTION

Squalene ($C_{30}H_{50}$) is a highly unsaturated isoprenoid hydrocarbon widely distributed in nature. Vegetable oils such as olive oil, palm oil, wheat germ oil, amaranth oil and rice bran oil contain varying amounts (0.1 to 0.7%) of squalene. It is also called as spinacene and supraene. Deep sea shark is the richest source of squalene in nature (Gopakumar, 1997). In fact, the compound got the name squalene as it was identified from the oil of the deep-sea shark belonging to the genus *Squalus*. The tropical deep-sea shark (*Centrophorus artomarginatus*) contains as high as 80% squalene in its liver oil. Basking shark liver oil also contains high amounts of squalene. Squalene is an all transisoprenoid with six isoprene units. Structurally it is called (all-E) 2, 6, 10, 15, 19, 23-hexamethyl-2, 6, 10, 14, 18, 22-tetrecosahexaene.

The ancient Shoguns of Japan were the first to recognize the beneficial effects of deep sea shark liver oil. They found that it provides strength, vigour, energy, virility and over all good health and called this very rare and precious extract as "Tokubetsu no Miyage" meaning "Special Gift". The Japanese fishermen of Suruga Bay in the Izu Peninsula, famous for shark fishing called this elixir as "Samedawa" meaning cure-all. The deep-sea shark liver also was in use by the coastal people and fishermen of Micronesia and they called it miracle oil. The Spanish mariners called it "aciete de bacalao" or the oil of the great fish. The Chinese ancient pharmaceutical book Honzokomuko also contains references of the therapeutic uses of deep sea shark liver oil. The credit for discovery of squalene was given to Dr. MitsumaruTsujimoto, a chemist on oils and fats at the Tokyo Industrial Testing Station who reported in 1906 that certain shark liver oils contain a highly unsaturated hydrocarbon and is rich in energy. The sharks belong to the family squalidae and the compound was named squalene. But it was after several years, in 1936, that the structural formula of squalene was elucidated by Paul Karrer, a Swiss Chemist working at Zurich University in Switzerland. Given the roles squalene performs, in the human body, it is aptly referred to as 'gift from sea' (Farvin *et al.*, 2009a).

1.2 SOURCES AND PROPERTIES OF SQUALENE

Deep sea sharks live about 900 m under the sea where sunlight and oxygen are almost negligible. Squalene is stored in the bodies of these sharks, which lack a swim bladder and therefore reduce their body density with fats and oils. Squalene, which is stored mainly in the shark's liver, is lighter than water with a specific gravity of 0.855. The ability of this species to withstand high pressure at this depth and to survive is due to squalene. Squalene abstracts oxygen from the water present in the body and releases it to the cells for physiological activities and also to provide strength and stamina.

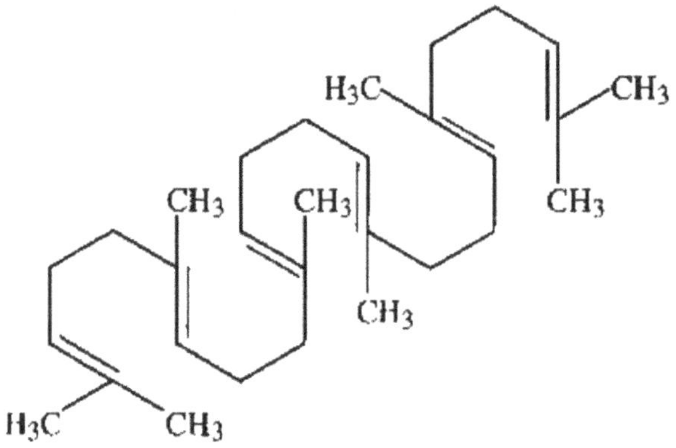

Fig. 1: Structure of Squalene

The richest amount of squalene is in the shark, *Centrophorus moluccensis* (synonym. *Centrophorus scalpratus*) abundantly occurring in the Indian Ocean, particularly in the seas of Andaman and Nicobar Islands. Gopakumar (1997) has reported that the liver oil of this species contained 70 per cent of squalene by weight. Extra virgin olive oil contains about 200-450 mg g^{-1} of squalene (Kelly, 1999). Extensive methodology for purification, estimation and industrial applications of squalene extracted from *C. scalpratus* has been reported (Gopakumar & Thankappan, 1986; Thankappan & Gopakumar, 1991). Characteristics of squalene are given in following table.

Table 1. Chemical properties of squalene

Properties	Values
Molecular weight	410.7
Melting point	-75°C
Viscosity at 25°C	12 centipoises
Specific gravity	0.8 to 0.86
Boiling point at 25°C	285°C
Calorific value	19 400 BTU Pound[-1]
Flash point	110°C

Squalene is present up to 85% by weight of liver in deep sea sharks. Among the plant sources, squalene is present in Amaranthus seed oil (6-8%), olive oil (up to 0.7% by weight), palm oil (0.1 to 2%, depending on species and method of extraction), rice bran oil and wheat germ (Liu *et al.*, 1976; Deprez *et al.*, 1990; Sun *et al.*, 1997; Newmark, 1997). Squalene extracted commercially from olive oil is marketed as vegetable squalene having purity of around 97.5% while squalene from shark liver oil can be processed up to 99.9% purity.

1.3 VEGETABLE OILS AS A SOURCE OF SQUALENE

Squalene is widely present as a component of the unsaponifiable fraction of vegetable oils.In general, the very low level of Squalene in vegetable oils does not represent a viable industrial source. In the case of Olive oil, high tonnages are physically refined. During this process the unsaponifiable fraction is concentrated in the refining condensate.

This provides the basis of an industrial source of Olive Squalene. Olive oil normally contains 0.2 – 0.4% of unsaponifiables. The unsaponifiables composition consists of squalene, natural olive waxes, tocophrol and phytosterol.

Squalene is obtained from by the process of hydrogenation of squalene at high pressure greater than 10 bars using nickel/nickel oxide as catalyst in an inert support. The reaction is highly exothermic.

Being a saturated hydrocarbon squalene is an excellent emollient oil with an exceptional silky touch. It hydrates and softens the skin rapidly. Squalene is rapidly absorbed into the epidermal layer and does not leave a greasy or sticky feeling. Squalene is compatible with most cosmetic ingredients and is stable over a wide range of temperatures. According to the Final Report of the American College of Toxicology (1982) both squalene and squalane are reported to be safe as cosmetic ingredients in concentrations currently used in trade.

Chemical Properties:

Molecular Formula	$C_{30} H_{62}$
Molar Mass	422.81g mol-^{1}, -38^{0}C,258K,-36^{0}F
Boiling point	176^{0}C at 0.05 mmHg.

Squalane: chemical structure

Saturated hydrocarbon C30H62

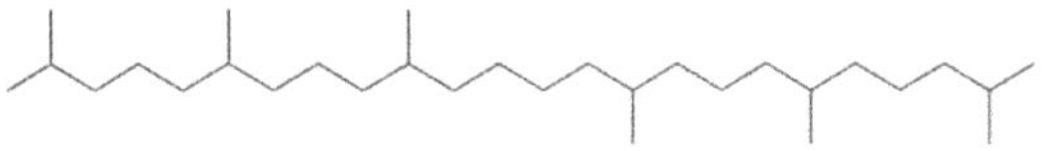

nci	Squalane
Iupac	2,6,10,15,19,23-Hexamethyltetracosane
Cas	111-01-3
Einecs	203-825-6

Fig. 2: Squalene: chemical structure

Squalene is commercially produced from two sources, from olive oil and deep sea shark liver oil.

1. **From deep sea shark liver oil:** Shark liver oil is kept frozen or under ice immediately after capture to prevent oxidative changes. The shark livers are cooked in a vat with stem heating facility from internally and externally. The oil separates at the top. The residue is pressed in a hydraulic press to get the last traces of oil. The oil is separated from water phase by decantation or centrifugation. The oil contains triglycerides, acylglyceryl ethers and free fatty acids. The free fatty acids, if present, are removed by refining (washing with dilute sodium bi carbonate solution). From this squalene is separated by vacuum distillation around 180^{0}C at 1mbar pressure. It can be further purified to get as white glassy liquid of 99.9% purity. Commercially squalene is produced from deep sea shark of Indian Ocean origin *Cetrophorus scalpratus* (photo below).

Fig. 3: Deep sea shark

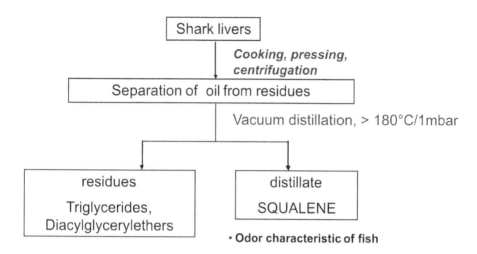

Fig. 4: Squalene – industrial process from shark liver oil

2. **Production process from olive oil:** The method of production from olive oil is an expensive process as natural olive oil is used as food and also several industrial applications. In olive oil it is present in the unsaponifiable matter to the extent of about 0.2 to 0.4% by weight of the unsaponifiables. The natural oil contains free fatty acids,diglycerides and triglycerides and unsuponifiable matter. The unsapofiable matter also contain traces of olive oil waxes, tocopherol, phytosterol, cholesterol etc.

In the industrial process of production of squalene first the free fatty acids are removes by refining. The refined oil is deodorized and dried free of moisture. It is then saponified to get the unsapoifable matter. From the unsaponified matter squalene is separated and purified.

Now-a-days olive squalene is more a preferred item for cosmetic preparations due to following reasons;

● It is a renewable resource compared to shark squalene.

● No fishy odour and vegetarian and non-fish-eating population prefer it compared shark squalene.

● Social and environmental issues as sharks are protected species and get being extinct.

● Olive squalene/vegetable squalene has greater stability to storage due presence other components like tocopherols present in it.

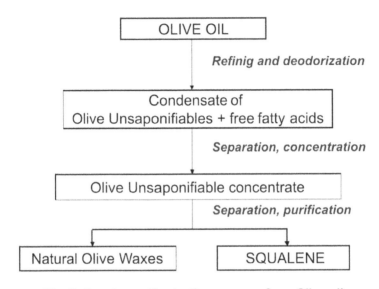

Fig. 5: Squalene - Production process from Olive oil

1.4 OCCURRENCE OF SQUALENE IN HUMANS

Squalene is a key intermediate in the biosynthesis of cholesterol. Over 60% of the ingested squalene is absorbed from the small intestine and then transported to the lymph in the form of chylomicrons into the circulatory system. In the blood, squalene combines with low density lipoproteins and is transported to various body tissues. A major portion of the absorbed squalene is distributed to the skin. Studies conducted on squalene in human adipose tissues show that fat tissues contain over 80% of the total squalene present and upto 10% in microsomal membranes (Koivisto & Miettinen, 1988; Stewart, 1992). Scientific evidence suggests that only microtonal membrane bound squalene (around 10% of total squalene) present in human body, is metabolically active and stimulates immune cells in the inner and outer coat of our body (Owen *et al.*, 2004). Squalene is present in important human tissues. The endogenous synthesis of squalene in animal body begins with the production of acetate from glucose. Acetate is converted to 3-hydroxyl-3-methylglutaryl coenzyme A (HMGCoA). The HMGCoA is reduced to mevalonate. The mevalonte is then phosphorylated in a three-stage process and then decarboxylated to form delta-3-isoprenyldiphosphate. This is followed by successive additions of prenyl groups with formation of 15-carbon farnesyl phosphate. Two molecules of farnesyl diphosphates are then enzymatically joined and reduced to form squalene by an enzyme called squalene synthetase (Kelly, 1999; Reddy & Couvreur, 2009). Various steps involved in the biosynthesis of squalene are given below (Fig. 6).

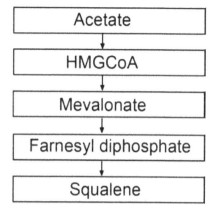

Fig. 6: Biosynthesis of squalene from acetate

Synthesis of cholesterol from squalene

Cholesterol is synthesized in human body from squalene. Squalene is cyclized by an enzyme called squalene cyclase to form cholesterol (Liu *et al.*, 1975). Steps involved in the process are cyclization and carbocation. Squalene epoxide formed is converted to lanosterol by the enzyme catalase. This is followed by the removal of 3 methyl groups from lanosterol to form cholesterol (Clayton & Bloch, 1956; Popják *et al.*, 1961) (Fig. 7).

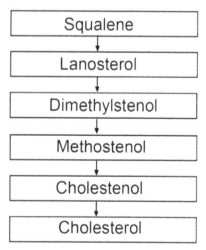

Fig. 7: Biosynthetic pathway of cholesterol from squalene

Synthesis of steroids from squalene

All steroids are derivatives of cholesterol. Cholesterol is hydroxylated by enzyme cytochrome P-450 (removes 6-member carbon side chain from cholesterol, to form pregnenolone, Clayton & Bloch, 1956; Popják *et al.*, 1961). This hormone is secreted by uterus to control ovum implantation. Pregnenolone is the precursor of androgens, estrogens and glucocorticoids. Testosterone is an androgenic steroid hormone produced mainly in the gonads and in smaller quantities by adrenal cortex.

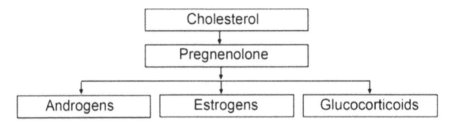

Fig. 8: Biosynthetic pathway of steroid hormones from cholesterol

Effect of squalene on cholesterol metabolism

Amounts of squalene present in human tissues vary, being rich in areas which have high sebaceous glands (face, forehead) and low in areas having poor sebaceous glands. Secretion of squalene by sebaceous glands of the skin amounts to 125-475 mg day^{-1} person^{-1} (Kelly, 1999). Although cholesterol is synthesized from squalene in human body, dietary intake of squalene does not elevate serum cholesterol level. Only microsomal bound squalene is actively involved in this process and rest of the squalene is stored in the lipid droplet (Tilvis & Miettinen, 1983). Strandberg (1990) has shown that when human volunteers were fed with squalene (90 mg day^{-1} person^{-1}) for 7-30 days, serum squalene levels increased up to 17-fold, but no significant changes were noticed in serum triglycerides and cholesterol contents. But squalene feeding produced significant increase in fecal excretion of cholesterol, its non-polar derivatives and bile salts suggesting that although cholesterol synthesis has increased by as much as 50%, faecal elimination was up-regulated. Because of high faecal elimination, no increase in serum cholesterol was noticed. Oral administration of squalene also results in the reduction of low density lipoprotein cholesterol (LDL), triglycerides and increase in high density lipoprotein (HDL).

1.5 SQUALENE AND CANCER

Olive oil is the staple source of dietary fat among people of Greece and many European countries. Epidemiological studies in the latter part of the 20th century showed that people who eat the so called Mediterranean diets (Gjonca & Bobak, 1997; Buckland *et al.*, 2011) have lower incidence of major disease like cancer and cardiovascular diseases. Consumption of olives and olive oil containing squalene is found to be responsible for these beneficial effects. Those women who consume large quantities of olive oil have lesser risk of breast cancer (Kelly, 1999; Newmark, 1997; Owen et al., 2004). It is now well established that olive oil contains squalene and the beneficial effect of resistance to cancer may be due to the squalene present in olive oil. The studies conducted on animals also support this observation of decreased incidence of breast cancer (Kate et al., 1992). Animal studies have shown that squalene is effective in inhibiting lung tumors. Squalene is also found to have chemo-preventive action against colon cancer and enhanced immune function (Nakagawa et al., 1985; Ikikawa et al., 1986; Rao et al., 1998). Hence, squalene is now being used as a prophylactic agent after irradiation treatment of cancer patients.

Squalene carries oxygen to cells (Strandberg, 1990; Wefers et al., 1991) and thereby enhances the function of several organs like liver and kidney (Kelly, 1999). It also prevents acidotic cell syndrome disease in which the cells become acidic and die due to absence of oxygen. Squalene is an excellent scavenger of free radicals and thereby protects cells and organs from auto-oxidative damages. Animal studies in recent times have shown that squalene in sebum played a protective role against formation of hydrocarbon carcinogens in skin and prevented damages caused by radiation therapy (Wefers et al., 1991; Strandberg, 1990). Being a natural product used by man from time immemorial, its consumption is harmless and has no side effects.

Cytoprotective activity of squalene

One of the major limitations of cancer chemotherapy is the indiscriminative injury of normal tissues, leading to multiple organ toxicity and consequent dose limitation/ treatment failure (Das et al., 2003). The resultant problems are myelo-suppression, renal toxicity and neuropathy having profound effects on adults and children with long term remission that affects not only therapy but also overall quality of life (Bukowski, 1999). Often, free radicals produced by chemicals and radiation therapy are the major problems. Scavenging these free radicals is important to minimize the toxicity of chemotherapy. Antioxidants are extensively used for this purpose. Several experimental studies conducted using squalene had demonstrated that it is an excellent detoxifying agent against toxic and carcinogenic chemicals like hexachlrobiphenyl, hexachlorobenzene, arsenic, theophilline, phenobarbital and strychnine (Kamimura et al., 1992; Richter & Schafer, 1982; Fan et al., 1996). Squalene has also been reported to have protective activity against nicotine-derived nitrosamineketone (NMK) -induced lung cancer (Smith et al., 1998).

Most antioxidants used today have side effects and thus fail to get approval from certifying agencies like the USFDA or WHO. Hence, there is an urgent need to develop new generation of antioxidants having essential characteristics like better safety profile, good oral availability and meeting basic qualities required for a cytoprotective agent (Das *et al.*, 2003). Squalene being a naturally occurring cytoprotective agent, toxicity of this compound has not been reported so far.

Squalene as antioxidant

Most healthy diets rich in good lipids contain squalene as a naturally occurring constituent (Rowland & Robson, 1986), especially one containing olive oil and deep-sea shark liver oil (Heller *et al.*, 1963). Squalene has unique ability to carry oxygen throughout the body without the help of haemoglobin. Being highly unsaturated, it contains 6 double bonds, and is extremely reactive to get into an oxidized state. Squalene extracts hydrogen from water and releases oxygen by the following reaction:

$$C_{30} H_{50} + 6 H_2O \longrightarrow C_{30} H_{62} + 3 O_2$$

The released oxygen enhances cellular metabolism giving vitality and vigour and also prevents acidosis of cell. Rajesh & Lakshmanan (2009) reported antioxidant effect of dietary squalene on sodium arsenite-induced oxidative stress in rat myocardium. Squalene is not very susceptible to peroxidation and appears to protect skin from lipid peroxidation due to exposure of UV radiation and other oxidative damage (Wefers *et al.*, 1991). Human body produces a number of peroxides generated by autoxidation of fat present in foods consumed. These peroxides which are free radicals produce a large quantity of carbonyl compounds harmful to human body. But cells possess their own defense mechanisms consisting of antioxidants. Antioxidants are very vital for health and they get oxidized at their expense and prevent damages by carbonyls. One of the antioxidant present in human body is vitamin E. Squalene is the best naturally occurring antioxidant which have no known side effects. A comparison of properties of squalene and vitamin E is given in Table 2.Low ionization threshold of squalene makes it an efficient donor of electrons without undergoing molecular disruption. This capacity makes squalene an excellent antioxidant. Oxidative damage of the skin starts at the surface induced by pro-oxidants which brings about oxidation of skin lipids (Ohsawa *et al.*, 1984; Wefers *et al.*, 1991). Due to its presence in skin squalene can effectively scavenge free radicals (Kohno *et al.*, 1995; Atkins, 2002; Das, 2005) from skin, one reason why squalene finds application as a skin moisturizer in cosmetics.

Table 2. Comparison of properties of squalene and vitamin E

Squalene	Vitamin E
A hydrocarbon, 30-carbon polyprenyl compound having 6 isoprene units.	An alcohol having 3 isoprene units called alpha- tocopherol (5,7,8-trimethyltocol).
Produced by human body, present in many tissues, rich in skin (McKenna, 1950). Squalene is available to human body both exogenously and endogenously.	Depend on dietary sources, has to be supplied through food.
Once it enters human body, squalene abstracts hydrogen from water in cells and releases oxygen to cells (Atkins, 2002).	Vitamin E has no such property
One of the strongest anti-oxidant found in nature (Kohno et al., 1995) and relatively stable to attack by free radicals as compared to Vitamin E	Powerful anti-oxidant but unstable to attack by free radical compared to squalene.
Increases male potency and vitality by increasing production of male hormones.	Has only the capacity to improve health
Strongly attracted to the hydrophobic bond between the two lipid layers of the biomembrane, where lipid per-oxidation is the greatest. Can move freely through the biomembrane without altering the properties of the biomembrane	Limited integration into the bio-membranes and cannot move freely
Can be used as an antidote to reduce drug toxicity(Kamimura et al., 1989)	Cannot be used
Squalene is stable after neutralization with free radicals and hence can be recycled.	Requires the use of other endogenous antioxidants like squalene for recycling
Have capacity to increase good cholesterol (HDL) and to reduce bad cholesterol (LDL).	No such properties are reported
Can tolerate uptake of up to 5g day-1 without any toxic effect.	No side effects at low dosages
Unlike other antioxidants like Lycopene, Vitamin A & E, squalene can be stored in high concentration in body (skin and adipose tissues)	Cannot be stored in high concentration

1.5.1 Squalene and Coronary Heart Disease (CHD)

The widespread occurrence of coronary heart disease is chiefly attributed to modern life style and food habits. Dietary squalene has been found to lower cholesterol levels in blood and also to reduce LDL and to increase HDL (Ikikawa *et al.*, 1986). This effect is due to efficiency of squalene to down-regulate the HMGCoA reductase which in turn enhances the capacity of liver to filter bad cholesterol. These findings are supported by epidemiological correlation studies of squalene rich olive oil consuming population having low incidence of CHD. The cholesterol lowering property of squalene has prompted pharmacologists to combine squalene with statin drugs (used to lower cholesterol levels) and to use in human therapeutic applications. This leads to lower doses of drug formulations andreduction in potential side effects of statins. This can also reduce in the long term, therapeutic costs of patients having hypercholesterolemia.

A clinical trial conducted by Chan *et al.* (1996) showed the effectiveness and safety of squalene alone and in combination with pravastatin in lowering the cholesterol levels. This double blind placebo-controlled, 20-week trial was conducted on a randomized selection of 102 elderly people, all suffering from high cholesterol levels. They received 10 mg pravastatin and/or 860 mg squalene daily either separately or in combination. The results showed that both pravastatin and squalene effectively reduced levels of total cholesterol and LDL cholesterol and increased levels of HDL cholesterol. The study concluded that (i) co-administration of pravastatin and squalene combined the effects of the two drugs on lipoprotein concentrations; (ii) the combination may be useful and cost-effective in elderly patients with hypercholesterolemia, who might have a higher incidence of side effects when using larger doses of pravastatin alone. Farvin *et al.* (2009b) reported the cardioprotective effect of squalene against isoprenaline-induced myocardial infarction in rats.

Although squalene is an intermediate in the cholesterol biosynthesis in animals, intake of squalene by man has not shown increase in serum cholesterol levels. On the contrary, squalene is seen to reduce cholesterol levels in man. This makes it safe for human consumption. In Greece and Italy, average intake of squalene is 200-400 mg day^{-1} while in United States, the consumption is only 30 mg day^{-1} (Yokota, 1997).

Effect of squalene on testosterone level, obesity and High Blood Pressure (HBP)

Obesity is a resilient and complex chronic disease. Leptin, an adipocyte-secreted hormone level has been shown to be responsible for obesity with several other hormones (Mantzoros, 1999). Leptin, a product of the *Lep* gene was discovered through the positional cloning technique used to determine the genetic defect resulting in obesity on ob/ob mice (Zhang *et al.*, 1994). The ob gene encodes a peptide from 167 amino acids named leptin (from Greek word *leptos*, means thin), whose crystal structure suggests that it belongs to cytokine family (Zhang *et al.*, 1997). This hormone has to be associated with weight loss, cachexia, inflammatory process, oxidative stress, arterial hypertension and aging. Another potential causative factor in obesity syndrome is leptin resistance (Mantzoros, 1999). Zhang et al. (2009) found that squalene regulates leptin, a compound controlling obesity in rats. Liu et al. (2009) found that feeding squalene may counteract increase in body fat, blood pressure, levels of plasma leptin, glucose, cholesterol, triglycerides and also effect increase in testicular weight and testosterone levels in rats. Similar results were seen in male red panda (Li *et al.*, 2003), boar (Zhang *et al.*,2008) and also in chicken (Li et al., 2010).

A comparative cross-sectional study conducted (Mendoza-Nunez et al., 2006) on 70 healthy elderly persons *viz.*, 46 women (mean age: 67±5.8) and 24 men (mean age: 73±7.5), and another group of 91 elderly persons with HBP composing 62 women (mean age: 67±8.2) and 29 men (mean age: 70±0.3) showed that elderly subjects with high blood pressure had significantly higher levels of leptin than healthy elderly subjects.

Squalene and skin care

Today, squalene is widely used in pharmaceutical formulations as an effective skin moisturizer and also to prevent formation of lines and wrinkles in skin associated with aging process or cancer. A variety of squalene incorporated cosmetic products are available in the markets all over the world. Besides being odourless and colourless, its properties such as high spreadability, light consistency, stability at all ambient temperatures, non greasy texture, rapid transdermal absorption, restoration of suppleness to skin, creation of moisture barrier, healing of chapped and cracked skin, antibacterial properties and ability to promote cell regeneration make squalene an excellent skin protector. Areas of application of squalene include anti-aging, wrinkle protection, eczema, damaged hair, dry scalp and brittle nail.

Squalene as adjuvant in vaccines

Immunological adjuvants are substances, administered in conjunction with vaccines that stimulate the immune system and increase response to vaccines (WHO,

2008). Squalene contact tests indicate that it is not a significant contact allergen or irritant and both squalene and its hydrogenated form, squalane are safe as cosmetic ingredients (IJT, 1982). MF59 is the propriety adjuvant name of squalene patented and in use by Novartis (Anon, 2009). This compound is added to human influenza vaccine to stimulate human body's immune response through production of CD4 memory cells. It is the first oil-in-water influenza vaccine commercially used. WHO and US Defense Department have both published extensive reports which emphasize that squalene is a chemical naturally occurring in human body (Asano *et al.*, 2002). According to the WHO, squalene has been present in over 22 million flu vaccines given to patients in Europe since 1997 and significant vaccine related adverse events were not reported (Roscoe, 2009).

Toxicity and dosages

Animal studies conducted have proved that there are no side effects or toxic signs in plasma, biochemical and hepatic functional tests (Kamimura *et al.*, 1989). Dosages depend on area of application, *viz.*, for lowering cholesterol, 860 mg day^{-1} was given along with pravastatin without any side effects. A dose of 500 mg day^{-1} was safe and had normalizing effect on lipid profile (Kelly, 1999). For cancer treatment in human beings, a dose of 2-5 g day^{-1} appears to be of therapeutic use. Since most olive oil consuming populations in the world, particularly in the Mediterranean area, who consume 500-800 mg squalene day^{-1}, have not experienced any ill effects for centuries, it can be concluded that a daily consumption of 500 mg squalene is safe and provide health benefits.

1.6 CONCLUSION

Available research findings have shown that substantial amount of dietary squalene is absorbed, a major part of which is used to synthesize cholesterol. This increase is not associated with elevation of plasma cholesterol in humans as a result of concomitant fecal elimination. Hence, the theory, that increased consumption of squalene is likely to enhance serum cholesterol is misplaced. A dietary supplementation of 500 mg of squalene day^{-1} appears to normalize plasma cholesterol levels, LDL and HDL cholesterol values. Squalene also confers many protective benefits to patients exposed to radiation treatment. Squalene assists in maintaining white cell counts during radiation treatment of cancer patients. Besides, it also protects the skin from many ill effects of UV radiation. The most exciting application of squalene for human health is its use as a safe and naturally occurring antioxidant. However, its availability in nature cannot meet the growing demand. The most abundantly available source of squalene is deep sea sharks which are not an inexhaustible resource. Another major source of squalene is olive oil but the content is very low. Besides, after extraction of squalene, the residual oil cannot be used for food purposes and hence is not an economically viable source. The need of the day is to synthesize squalene by organic hemisynthesis.

1.7 REFERENCES

1. Anon (2009) MF59 Adjuvant Fact Sheet, Novartis, June 2009. www.novartis.com.ph/downloads/newsroom/MF59-Adj-fact-sheet.pdf (Accessed 12 November 2011).

2. Asano, K.G., Bayne, C. K., Horsman, K.M. and Buchanan, M.U. (2002) Chemical composition of finger prints for gender determination. J. Forensic Sci. 47(4): 805-807.

3. Atkins, R.C. (2002) Squalene: Oxygenator, Cancer fighter. Dr Atkins' Vita-Nutrient Solution- Nature's Answer to Drugs, UK Pocket Books, Simon & Schuster, UK.

4. Buckland, J.G., Agudo, A., Noemie Travier, N., Huerta, J.M., Cirera, L., Torm, M.J., Navarro, C., Chirlaque, M.D., Moreno-Iribas, C., Ardanaz, E., Barricarte, A., Etxeberria, J., Marin, P., Quirãs, J., Redondo, M. L., Larraãaga, N., Amiano, P., Dorronsoro, M., Arriola, L., Basterretxea, M., Sanchez, M.J., Molina, E. and Gonzãlez, C.A. (2011) Adherence to the Mediterranean diet reduces mortality in the Spanish cohort of the European prospective ivestigation into cancer and nutrition (EPIC-Spain). British J. Nutrition. 106 (10): 1581-1591.

5. Bukowski, R. (1999) The need for cytoprotection. Eur. J. Cancer. 32A: S2-S4.

6. Chan, P., Tomilsonm, B., Leemm, C.B. and Lee, Y.S. (1996) Effect and safety of low dose prevastatin and squalene alone in combination in elderly patients with hypercholesterolemia. J. Clin. Pharmacol. 36: 422-427.

7. Clayton, R.B. and Bloch, K. (1956) The biological conversion of lanosterol to cholesterol. J. Biol. Chem. 218: 319.

8. Das, B. (2005) The Science Behind Sqalene -The Human Antioxidant, 2ndedn., Toronto Medical Publishing, Canada.

9. Das, B., Yeger, H., Baruchel, H., Freedman, M.H., Koren, G. and Baruchei, S. (2003) In vitro cytoprotective activity of squalene on a bone marrow

versus neuroblastoma model of cisplatin-induced toxicity, implications in cancer chemotherapy. Eur. J. Cancer. 39(17): 2556-2565.

10. Deprez, P.P., Volkman, J.K. and Davenport, S.R. (1990) Squalene content and neutral lipid composition of livers from deep-sea sharks caught in Tasmanian Waters. Aust. J. Mar. Fresh. Res. 41: 375-387.

11. Fan S., Ho, I., Yeoh, F.L., Lin, C. and Lee, T. (1996) Squalene inhibits sodium arsenite-induced sister chromatid exchanges and micronuclei in Chinese hamster ovary-K1 cells. Mut. Res. 368: 165-169.

12. Farvin, K.H.S., Anandan, R., Kumar, S.H.S., Sussela, M., Sankar, T.V. and Nair P.G.V. (2009b) Biochemical studies on the cardioprotective effect of squalene against isoprenaline-induced myocardial infarction in rats. Fish. Technol. 46: 139-150.

13. Farvin, K.H.S., Raj, A.S., Anandan, R. and Sankar, T.V. (2009a) Squalene-A gift from sea.Sci. India. 12: 20-24.

14. Gjonca, A. and Bobak, M. (1997) Albanian paradox- Another example of protective effect of Mediterranean life style.Lancet. 350: 1815-17.

15. Gopakumar, K. (1997) Tropical Fishery Products, 190 p, Oxford & IBH Publishing Co. Pvt. Ltd., New Delhi.

16. Gopakumar, K. and Thankappan, T.K. (1986) Squalene- Its sources, uses, and industrial applications. Seafood Export J. 18(3): 17-21.

17. Heller, J.H., Pasternak, V.Z., Ransom, J.P. and Heller, M.S. (1963) A new reticulo-endothelial stimulating agent from shark livers.Nature. 199: 904-905.

18. IJT (1982) Final report on the safety assessment of squalane and squalene. Int. J. Toxicol. 1: 37-56.

19. Ikikawa, T., Umeji, M., Manabe,T., Yanoma, S., Orinoda, K. and Mizunuma, H. (1986) Studies on antitumor activity of squalene and its related compounds (Japanese). Yakugaku Zasshi. J. Pharmac. Soc. of Japan. 106: 578-582.

20. Kamimura, H., Koga, N., Oguri K., Yoshimura, H., Inoue, H., Sato, K. and Ohkubo, M. (1989) Studies on distribution, excretion and subacute toxicity of squalene in dogs. Fukuoka Igaku Zasshi, 80: 269-280.

21. Kamimura, H., Koga, N., Oguri, K. and Yoshimura, H. (1992) Enhanced elimination of theophylline, phenobarbital and strychinine from the bodies of rats and mice by squalene treatment. J. Pharmacobiodyn. 15: 215-221.

22. Kate, K., Cox, A.D., Hisaka, M.M., Graham, S.M., Buss, J.E. and Der, C.J. (1992) Isoprenoid addition to Ras protein is the critical modification for its membrane association and transforming activity. Proc. Natl. Acad. Sci. U. S. A. 89: 6403–6407.

23. Kelly, G.S. (1999) Squalene and its potential clinical uses.Altern. Med. Rev. 4(1): 29-36.

24. Kohno, Y., Egawa, Y., Itoh, S., Nagaoka, S., Takahashi, M. and Mukai, K. (1995) Kinetic study of quenching reaction of singlet oxygen and scavenging reaction of free radical by squalene in n-butanol. Biochim.Biophys.Acta. 1256(1): 52-56.

25. Koivisto, P.V.I. and Miettinen, T.A. (1988) Increased amount of cholesterol precursors in lipoproteins after ileal exclusion. Lipids. 23: 993-996.

26. Li, C., Wei, F., Li, M., Liu, X., Yang, Z. and Hu, J. (2003) Fecal testosterone levels and reproductive cycle in male red panda (Ailurus fulgens).Acta Theriolgica Sinica. 23(2): 115-119.

27. Li, S., Liang, Z., Wang, C., Feng, Y., Peng, X. and Gong, Y. (2010) Improvement of reproduction of performance I AA+ meat-Type male chicken by feeding with squalene. J. Anim. Vet. Adv.9(3): 486-490.

28. Liu, C.K., Ahrens Jr., E.H., Schreibman, H. and Crouse, R. (1976) Measurement of squalene in human tiss.ues and plasma: validation and application. J. Lipid Res. 17: 38–45

29. Liu, G.C.K., Ahrens Jr, E.H., Schreibman, P.H., Samuel, P., McNamara, D.J. and Crouse, J.R. (1975) Mechanism of cholesterol synthesis in man by isotope kinetics of squalene. Proc. Natl. Acad. Sci. USA. 72: 4612–4616.

30. Liu, Y., Xu, X., Bi, D., Wang, X., Zhang, X., Dai, H., Chen, S. and Zhang, W. (2009) Influence of squalene feeding on plasma leptin, testosterone & blood pressure in rats. Indian J. Med. Res. 129: 150-153.

31. Mantzoros, C.S. (1999) The role of leptin in human obesity and disease: a review of current evidence. Ann. Intern. Med. 130(8): 671-680.

32. McKenna, R.M.B. (1950) The composition of surface skin fat (sebum) from the human forearm. J. Invest Dermatol. 15: 33-47.

33. Mendoza-Nunez, V.M, Correa-Munoz, E. and Garfias-Cruz, E. A. (2006) Hyperleptinemia as a risk factor for high blood pressure in the elderly. Arch. Pathol. Lab. Med. 130: 170-175.

34. Nakagawa, M., Yamaguchi, T., Fukawa, H., Ogata, J., Komiyama, S., Akiyama, S. and Kuwano, M. (1985) Potentiation by squalene of the cytotoxicity of anticancer agents against culture mammalian cells and murine tumor. Jpn. J. Cancer Res. 76: 315-320.

35. Newmark, H.L. (1997) Squalene, olive oil, and cancer risks: a review and hypothesis. Cancer Epidemiol. Biomarkers Prev. 6: 1101-1103.

36. Ohsawa, K., Watanabe, T., Matsukawa, R., Yoshimura, Y. and Imaeda, K. (1984) The possible role of squalene and its peroxide of the sebum in the occurrence of sunburn and protection from the damage caused by U.V. irradiation. J. Toxicol. Sci. 9(2): 151-159.

37. Owen, R.W., Haubner, R., Wurtele, G., Hull, E., Spiegelhalder, B. and Bartsch, H. (2004) Olives and olive oil in cancer prevention. Eur. J Cancer Prev. 1394: 319-326.

38. Popják, G., Goodman, W.S., Cornforth, J.W., Cornforth, R.H. and Ryhage, R. (1961) Studies on the biosynthesis of cholesterol. XV. Mechanism of squalene biosynthesis from farnesyl pyrophosphate and from mevalonate. J. Biol. Chem. 236: 1934-1947.

39. Rajesh, R. and Lakshmanan, P.T. (2009) Antioxidant defense of dietary squalene supplementation on sodium arsenite-induced oxidative stress in rat myocardium.Int. J. Biomed. Pharm. Sci. 2: 98-108.

40. Rao, C.V., Newmark, H.L. and Reddy, B.S. (1998) Chemopreventive effect of squalene on colon cancer.Carcinogenesis. 19: 287-290.

41. Reddy, L.H. and Couvreur, P. (2009) Squalene: A natural triterpene for use in disease management and therapy. Advanced Drug Delivery Rev. 61: 1412–1426.

42. Richter, E. and Schafer, S.G. (1982) Effect of squalane on hexachlorobenzene (HCB) concentrations in tissues of mice. J. Environ. Sci. Health. 17: 195-203.

43. Roscoe, W. (2009). The biology of influenza and vaccination. Issues4science's webloghttp://issues4science.wordpress.com/2009/11/07/the-biology-of-influenza-and-vaccination/ (Accessed 03 November 2011).

44. Rowland, S.J. and Robson, N.J. (1986) Identification of novel widely distributed sedimentary acyclic sesterpenoids. Nature. 324: 561-563.

45. Smith, T.J., Yang, G.Y., Seril, D.N., Liao, J. and Kim, S. (1998) Inhibition of 4-(methylnitrosamino)-1-(3-pyridyl)-1-butanone-induced lung tumorigenesis by dietary olive oil and squalene. Carcinogenesis. 19: 703-706.

46. Stewart, M.E. (1992) Sebaceous gland lipids. Semin.Dermatol. 11: 100-105.

47. Strandberg, T.E. (1990) Metabolic variables of cholesterol during squalene feeding in humans: comparison with cholestyramine treatment. J. Lipid Res. 31: 1637-1643.

48. Sun, H., Wiesenborn, D., Tostenson, K., Gillespie, J., and Rayas-Duarte, P. (1997) Fractionation of squalene from amaranth seed oil. J. Am. Oil Chem. Soc. 74, 413-418.

49. Thankappan, T.K. and Gopakumar, K. (1991) A rapid method of separation and estimation of squalene from fish liver oil using Iatroscan analyser. Fish. Technol. 28: 63-66.

50. Tilvis, R.S. and Miettinen, T.A. (1983) Absorption and metabolic fate of dietary 3H-squalene in rat.Lipids. 18 (3): 233-238.

51. Wefers, H., Melnik, B.C., Flür, M., Bluhm, C., Lehmann, P. and Plewig, G. (1991) Influence of UV irradiation on the composition of human stratum corneum lipids.J. Invest. Dermatol. 96: 959-962.

52. WHO (2008) Global Advisory Committee Report on Vaccine Safety,http://www.who.int/vaccine_safety/topics/adjuvants/squalene/questions_and_answers/en/ (Accessed 07 January 2012).

53. Yokota, T. (1997) Squalene - Treasure of the Deep. Yokota Health Institute, Tokyo, Japan.

54. Zhang, F., Basinski, M.B., Beals, J.M., Briggs, S.L., Churgay, L.M., Clawson, D.K., DiMarchi, R.D., Furman, T.C., Hale, J.E., Hsiung, H.M., Schoner, B.E., Smith, D.P., Zhang, X.Y., Wery, J.P. and Schevitz, R.W. (1997) Crystal structure of the obese protein leptin-E100. Nature. 387: 206-209.

55. Zhang, Q., Yang, S., Yang, L. and Zhang, W. (2009) Leptin concentration and breeding and body fat of rat in response to feeding squalene. J. Huzhong Agricultural University. 28(1): 58-60.

56. Zhang, W., Zhang, X. and Chen, S. (2008) Feeding with supplemental squalene enhances the productive performances in boar. Anim. Reprodu. Sci. 104 (24): 445-449.

57. Zhang, Y.Y., Proenca, R., Maffei, M., Barone, M., Leopold, L. and Friedman, J. M. (1994) Positional cloning of the mouse obese gene and its human homolog. Nature. 372: 425-432.

CHAPTER **2**

SHARK CARTILAGE

2.1 HISTORICAL

Shark cartilage bone has been used in traditional medicine in many Asian countries for centuries. It is used to treat cancer, particularly for lung cancer. It is used in the form of powder or soluble extract. This belief is based on an assumption that shark never get cancer. It was William Lanes Brok in 1922 first reported that shark does not get cancer. This has created a scientific controversy. Now it is proved that shark also get occasionally cancer mostly in soft tissues. But this statement created an opportunity for many shark bone processors to sell the shark cartilage as anti-cancer medicine.

People use today many of the parts of shark body like meat, oil and bones in particular for treatment of cancer. It is propounded that chondroitin stops the formation of blood vessels that supply oxygen and other nutrients to cancer cells. Another theory popular is that shark cartilage components promote immune system. It is also believed that certain ingredients in shark bones contain cancer inhibiting compounds. In recent years several clinical trials have been conducted on the use of shark cartilage on its effect on cancer both in animals and man. These studies conclusively proved that this is a pure myth and so-called claims of anti-cancer properties of shark cartilage are false. However, shark cartilage bone powder and extracts still enjoy a lucrative market in traditional medicine globally.

There are many naturally occurring compounds in both plants and animals that can fight against many diseases. Shark bone is one such compound used in chemotherapy as a neutraceutical.

Preparation

Shark cartilage bones of all shark species are used to prepare the powder. The entire exoskeleton of shark is composed of cartilage. In terms of weight the cartilage content comes up to 6% w/by body wt. Hence the quantity of shark cartilage that can be obtained from a single shark is quite high. The bones are washed thoroughly in clean water. Then these bones are washed in mild alkali to remove the adhering proteins. The washed cleaned bones are the treated with hydrogen peroxide to bleach the bones to get white colour. These bones are sun dried and packed for sale. Bones are powdered and sieved to get a fine powder. Many of the medical properties of shark cartilage may be attributed to chondroitin. Composition of chondroitin in given below (Table 3):

Table 3: Composition of Chondroitin

Characteristics	Composition
Colour	White powder with a characteristic smell.
pH	5.5- 7.5
Proteins	3-4%
Chlorides	2-4%
Heavy metals	20-50 ppm depending on species and place of capture
Principal constituent.	Chondroitin sulphate, up to 90% by dry weight basis

Chondroitin

It exists in natural form as a sulphate and is an essential component of all human and animal cartilages belonging to a family of complex polysaccharides called glucosaminoglycans (GAGs). In cartilage tissues GAGs account for 5-20% of the total components. Chondroitin occurs in connective tissues as a sulphated compound of repeating disaccharide units, Chondroitin has been used for several years for treatment of several diseases in man particularly for arthritis, psoriasis and cancer. In recent times chondroitin has been used in conjunction with glucosamine to reduce joint pain caused by osteoarthritis and to prevent cartilage loss associated with this disease. Structurally chondroitin is very similar to heparin.

Chondroitin sulphate belongs to a family of complex polysaccharides called glucosaminoglycans(GAGs). Chemically it is composed of long chains of repeating units of disaccharides called chondrosines. These chondrosines are composed of D- glucuronic acid and D- galactosamine. Depending one site of attachment of these chondrosines each other they are classified. Based on this there are mainly two types of chondroitins naturally occurring. They are Chondroitin A(4-sulthate) and Chondroitin C (6- sulphate). Commercially available chondroitin is a mixture of both A&C. The C form is a usually a minor one. Structure of chondroitin is similar to heparin (Fig. 9).

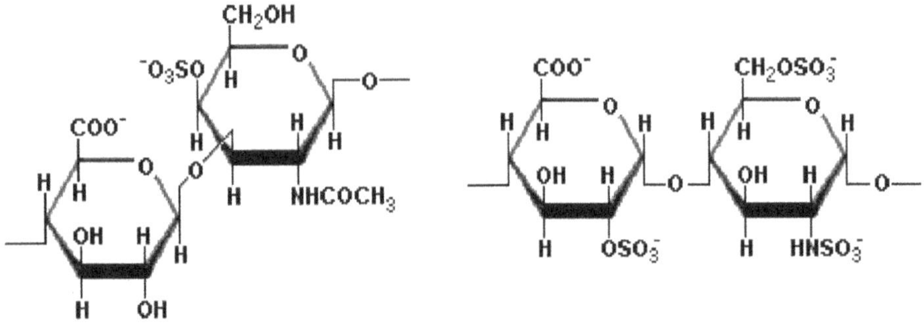

Chondroitin Sulfate   **Heparin**

Fig. 9: Structure of Chondroitin and Heparin

Cartilage collagen in higher vertebrates conforms to Type I collagen. But type I collagen accounts only one third of the total collagen content of shark cartilage(Rama and Chandrakasan, 1984). Shark cartilage chondroitin sulphate (Mr 48000,average repeating disaccharide units o.4%) was found consisting of following composition (Uchiyama *et al.*, 1987).

- Chondroitin 6- sulphate   66.8%
- Chondroitin4-sulphate   22.5%
- Chondroitin disulphated(D type)   10.3 % and,
- Chondroitin nonsulphated   0.4%

The polysaccharides isolated from chondroitin sulphate of shark cartilage were found to have high degree of structural diversity reflecting the enormous heterogeneity of parent polysaccharide. Analysis of the different sequence hydrolysates, as many as of 24 of them, revealed that 19 of them were novel ones (Deepa *et al.*, 2007).

Biosynthesis of chondroitin sulphate (Sibert and Sugumaram, 2002)

Chondroitin sulphate / dermatin sulphate are synthesized as galactosaminoglycan polymers containing N-acetylgalactosamine alternating with glucuronic acid. The sugar residues present are seen sulphated to varying degrees and positions depending on the tissue sources and varying conditions of formation. Epimerization of any the glucuronic acid residues to iduronic acid at the polymer level constitutes the formation of dematin sulphate. Both chondroitin/dermatin glycosaminoglycans are covalently attached by a common tetrasaccharide sequence to serine residues of core proteins while they are adherent to the inner surface of endoplasmic reticulam. This is followed by the addition of two galactose residues by two distinct glycosyl transferase enzymes in early cis/ medial regions of the Golgi. The linkage of tetrasaccharide is completed in the medial/ trans Golgi by the addition of the first glucuronic acid residue, followed by transfer of N-acetyl galactosamine to initiate the formation of a galactosaminoglycan rather than glucosaminoglycan. According to the above scientists this specific enzyme, N-acetylgalactosaminyl transferase, is different from the enzyme chondroitin synthase involved in the generation of the repeating disaccharide units to form chondroitin polymer. Sulphation of chondroitin polymer is done by another specific enzyme called sulfotransferase.

2.2 SHARK CARTILAGE AND OSTEOARTHRITIS

There are two principal activities naturally occurring and attributed to shark cartilage in conventional chemotherapy. They are anti-inflammatory and chondroprotective functions. The glycosaminoglycan present in shark cartilage chondroitin inhibits cartilage degradation. This ability is due to the efficiency of glycosaminogycans to inhibit the specific enzyme responsible for the degradation process of the cartilage. These enzymes belong to the group of enzymes called metalloproteinases and elastases. In a normal healthy person this is not a problem. But once arthritis sets in the joints the enzymes become more active and the process of degradation of collagen in the joints takes place very rapidly.

It is a form of arthritis caused by wearing a way of or degeneration of the cartilage that cushions the end of bones. This disorder is very common among elder people particularly in the knees. Drying of cartilage tissues in osteoarthritis is a major cause of tissue destruction. Tissue fluid keeps cartilage healthy in two ways. It acts as a shock absorber within the joints of the body, thus protecting the cartilage in the joints from being worn away. The cartilage in both sharks and human body have no blood vessels to supply the nutrients. It essentially gets the nutrients from the fluid. Drying of the fluids therefore caused the destruction of the tissues. Chondroitin when given helps to attract and hold fluid within the cartilage tissues. Glycosaminoglycans absorb free radicals to prevent further oxidative damages to cartilage tissues. Another important action of shark cartilage is the inhibition of *angiogenesis* (birth of new bloodvessels, Lee *et al.*, 1983.; Davis *et al.*, 1997). Although *angiogenesis* is an essential body function, uncontrolled capillary growth seen in rheumatoid arthritis is responsible for the rapid destruction of joints.Growing tumors require angiogenisis to bring them more oxygen, sugar and nutrients. Without blood supply, tumors cannot grow to abnormal size as is seen in cancer. Blood vessels also serve as a high way to cancer cells to move to other parts of the body. This process is clinically called metastazis. Most drugs used to combat cancer are inhibitors of angiogenisis.

Shark cartilage by inhibiting *angiogenesis* also reduces this cartilage - destructive process. Recent research investigations of Imada et.al (2010) show that shark cartilage chondroitin sulphate (CS) and porcine trachea cartilage (CS-PC) composed of mostly chondroitin 4- sulphate are likely to be multifunctional chondrioprotective material for treatment of degenerative arthritis disease as they can effectively suppress the activities of both aggrecanase-2 and metalloproteinase activities. This is a very significant research investigation supporting the action of shark cartilage in the prevention of rheumatoid arthritis.

Shark cartilage and cancer

In the conventional Chinese medicine shark cartilage is used to treat cancer. It is available all over the world to be used only as a dietary supplement (Holt,1995). But people are using it in traditional medicine for treatment of cancer. But its effectiveness in treatment of any type of cancer is not supported by clinical studies or competent agencies like WHO or US FDA. The reports on its effectiveness as an anti-cancer drug are also contradictory.

The use of shark cartilage or anybody part of the body of shark was based on a belief that sharks do not get attacked by cancer. But this was proved to be a mistaken belief. Extensive investigations done by George Washington University Medical Center reported that over thirty tumors were detected in elasmobranches those includes sharks also (Harshbarger, 1999). This report proves that the belief sharks do not get cancer is only a myth. However, the report of Prieur *et al.* (1976) shows that sharks have a reduced level of cancer. This view is interpreted by many scientists that sharks have some form of resistance to the development of cancer. Perhaps this may be the reason why shark cartilage is still used by many for treatment of cancer. But more research is needed to confirm these findings.

2.3 CLINICAL STUDIES ON HUMAN

Studies conducted on animals using shark cartilage for cancer treatment partly support the beneficial effects. Oral administration of shark cartilage was found to inhibit bFGF-induced angiogenesis as did oral administration of thalidomide in *in vivo* model studies on rabbits (Gonzales *et al.*, 2001). But so far, all studies conducted on human beings are disappointing. Study using oral administration of shark cartilage powder for 20 weeks on 20 prostate/ breast cancer patients (Leitner *et al.*,1983), shark collagen orally on different cancer patients for 6 weeks (Miller *et al.*, 1998) and using a shark cartilage extract, trade name AE-941, a Phase III clinical trial on cell lung cancer (Lu et al., 2007) and another Phase III trial on renal carcinoma (Escudier *et al.*, 2007) were all without much success.

Clinical studies by Mayo Clinic USA

One of the important studies conducted in USA by Mayo Clinic is worth recording. This is a study of two arm, randomized placebo-controlled, double-blind clinical trial. Patients with incurable breastor colorectal carcinoma had to have good performance status and organ function. Patients could be receiving chemotherapy. Patients were also to receive a shark cartilage product or an identical appearing and smelling placebo 3-4 times a day.

Result: 83 evaluable patients were analyzed. The trial was unable to demonstrate any suggestion of the efficiency of this shark cartilage product in patients with advanced cancer.

Some general uses of shark cartilage/chondroitin sulphate

Chondroitin is also thought to protect cartilage by following actions:

- Anti- inflammatory activity by inhibiting Cyclo-oxygenaseII activity.
- Inhibiting the activity of enzymes responsible for cartilage break down.
- Counteracting enzymes that interfere with the transport of nutrients to the cartilage.
- Stimulates the production of proteoglycans, glycosaminoglycans, and collagen. These complex molecules are the building block of new cartilage.

Chondroitin sulphate is used for following medical applications:

- Neural headache
- Neuralgia
- Arthritis
- Hemicarnia
- Reducing blood lipid
- Used in food preparations as health supplement.

Side effects and precautions of using shark cartilage

1. Shark cartilage contains high amounts of calcium. Prolonged usage can result in formation of kidney stones.

2. Shark cartilage should not be taken by people recovering from invasive surgery, pregnant women and people undergoing wound healing. This is because in such cases angiogenesis should not be inhibited.

3. Shark cartilage products should not be taken along with fruits and vegetables as they contain flavanoids. Bioflavanoids are found to get attached with proteins. The angiogensis - inhibiting property of shark cartilage is due to a protein present in it. Hence removal of this protein can result in the loss of this property. Hence it is advised to take shark cartilage product at least two hours before diet containing fruits and vegetables.

2.4 CONCLUSIONS

There are positive evidences to conclude that using shark cartilage is beneficial for treatment of arthritis. But with regard to use as a therapeutic agent for cancer, the results are not supporting. The clinical studies using cancer people have several limitations to get results. Patients with advanced stage of cancer the shark cartilage may not give much result as the patient is already using powerful drugs and shark cartilage is always given as additionality. Hence its impact may not be fully applicable. But people in initial stage of cancer and not using any anti-cancer drug may get some relief. This is one reason people still use this product as anti -cancer drug despite not getting any support from clinical research. There is need to conduct more research in the use of shark cartilage for application in cancer treatment.

2.5 REFERENCES

1. D. J. Prieur, J. K. Fenstermacher, A. M. Guarino, J. Natl. Cancer Inst. 56. 1207 (1976); S. R. Wellings, Natl. Cancer Inst. Monogr. 31, (1969), p. 59/ J. C. Harshbarger, Activities Report of The Registry of Tumors in Lower Animals, 1965-1973 (Smithsonian Institution, Washington, D.C., 1974) [http:/ /www.luogocomune.net/site/modules/mydownloads/library/acrobat/1185.pdf].

2. Davis PF, He Y, Furneaux RH, Johnston PS, Rüger B and Mand Slim GC. Inhibition of angiogenesis by oral ingestion of powdered shark cartilage in a rat model. Microvasc Res. 1997 Sep;54(2):178-82.

3. Deepa SS, Yamada S,Fukin S,Sugahara K., Structural determination of novel
    sulfated octasaccharides isolated from chondroitin sulfate of shark cartilage
    and their application for characterizing monoclonal antibody epitopes.
    Glycobiology. 2007, Jun. 1796) 631-45, Eupb. 2007 Feb. 22.

4. Den Boer JA and Westenberg HG. Behavioral, neuroendocrine, and biochemical
    effects of 5-hydroxytryptophan administration in panic disorder. Psychiatry
    Res. 1990 Mar;31(3):267-78.

5. Diego Andolina, David Conversi, Simona Cabib,Antonio Trabalza,2 Rossella
    Ventura, Stefano Puglisi-Allegra,1,2 and Tiziana Pascucci1, 2 5-
    Hydroxytryptophan during critical postnatal period improves cognitive
    performances and promotes dendritic spine maturation in genetic mouse model
    of phenylketonuria. Int J Neuropsychopharmacol. 2011 May; 14(4): 479–
    489.Published online 2010 Nov 1. doi: 10.1017/S1461145710001288.

6. Escudier B, Choueiri TK, Oudard S, Szczylik C., Négrier S , Ravaud A ,
    Chevreau C , Venner P , Champagne P , Croteau D , Dupont E , Hariton
    C and Bukowski RM Prognostic factors of metastatic renal cell carcinoma
    after failure of immunotherapy: new paradigm from a large phase III trial
    with shark cartilage extract AE 941. J Urol 178 (5): 1901-5, 2007.Gonzales,
    et al. Demonstration of inhibitory effect of oral shark cartilage on basic
    fibroblast growth factor-induced angiogenesis in rabbit cornea., Biol. Pharm.
    Bull. 2001, Feb (2), 151-4.

7. González RP1, Soares FS, Farias RF, Pessoa C, Leyva A, de Barros Viana
    G and, Moraes MO. Demonstration of inhibitory effect of oral shark cartilage
    on basic fibroblast growth factor-induced angiogenesis in the rabbit cornea.
    Biol Pharm Bull. 2001 Feb;24(2):151-4.

8. Harshbarger JC. 1999 (pers. comm.). Registry of Tumors in Lower Animals.
    Department of Pathology, George Washington University Medical Center,
    Washington, D.C. [http://www.vin.com/Proceedings/Proceedings.plx?CID=
    IAAAM1999&Category=&PID=49514&O=Generic]

9. Holt S: Shark cartilage and nutriceutical update. Altern Complement Ther 1:
    414-16, 1995. [http://cancernet.nci.nih.gov/cancertopics/pdq/cam/cartilage/
    HealthProfessional/page3].

10. Imada K., Oka H, Kawasaki D,Miura N, Sato T, Ito A. Anti- arthritic action
    mechanism of natural chondroitin sulphate in human articular chodrochytes
    and synovial fibroblasts. Boil. Pharm. Bull. 2010.33(3) 410-4.

11. Lee, Anne, and Robert Langer. "Shark Cartilage Contains Inhibitors of Tumor Angiogenesis." Science 221 (Sep. 16 1983): 1185-187. [PUBMED].

12. Leitner SP, Rothkopf MM, Haverstick L, *et al.*: Two phase II studies of oral dry shark cartilage powder (SCP) with either metastatic breast or prostate cancer refractory to standard treatment. [Abstract] Proceedings of the American Society of Clinical Oncology 17: A-240, 1998. [http://www.asco.org/ASCOv2/Meetings/Abstracts?&vmview=abst_detail_view&confID=31&abstractID=12633].

13. Loprinzi CL, Levitt R, Barton DL, Sloan JA, Atherton PJ, Smith DJ, Dakhil SR, Moore DF Jr, Krook JE, Rowland KM Jr, Mazurczak MA, Berg AR and Kim GP; North Central Cancer Treatment Group.Psychiatry Res. 1990 Mar;31(3):267-78.

14. Loprinzi CL, Levitt R, Barton DL, Sloan JA, Atherton PJ, Smith DJ, Dakhil SR, Moore DF Jr, Krook JE, Rowland KM Jr, Mazurczak MA, Berg AR and Kim GP; Evaluation of shark cartilage in patients with advanced cancer: a North Central Cancer Treatment Group trial. Cancer. 2005 Jul 1;104(1):176-82.

15. Lu C, Lee JJ, Komaki R, *et al.*: A phase III study of AE-941 with induction chemotherapy (IC) and concomitant chemoradiotherapy (CRT) for stage III non- small cell lung cancer (NSCLC) (NCI T99-0046, RTOG 02-70, MDA 99-303). [Abstract] J Clin Oncol 25 (Suppl 18): A-7527, 391s, 2007. [http://www.asco.org/ASCOv2/Meetings/Abstracts?&vmview=abst_detail_view&confID=34&abstractID=31161].

16. M. L. Moss, Am. Zool. 17, 335 (1977) [http://www.luogocomune.net/site/modules/mydownloads/library/acrobat/1185.pdf].

17. Miller DR, Anderson GT, Stark JJ, Granick JJ, Richardson D. Phase I/II trial of the safety and efficacy of shark cartilage in the treatment of advanced cancer. J Clin Oncol 1998;16(11):3649-55.

18. Rama S,Chandrakasan G. Distribution of different molecular species of collagen in vertebral cartilage of shark ((Carcharius acutus).Cunnect. Tissue Res. 1984, 12(2) 111-8.

19. Silbert, J.E., Sugumaram. Biosynthesis of Chondroitin/dermatin sulphate. IUBMB- Life, 2002, Oct. 54(4) 177-86.

20. Uchiyama H, Kikuchi,K., Ogama A, Nagasawa K. determination of the distribution of constituent disaccharide units within the chain near linkage region of shark cartilage chondroitin C. Biochim Biophys Acta 1987. Dec. 7, 926 93) 239-48.

CHAPTER **3**

OYSTERS

3.1 HISTORICAL

Oysters have been existing in the world for more than 180 million years ago. They are one of the oldest marine organisms living today. There are historical evidences showing that it was one of the staple food consumed by Neolithic man, lived on earth five thousand years ago, in large quantities. The word "*ostracise*" was derived from the Greek word "*astrakeon*". Aristotle in 320 B. C. investigated the regenerative process of oysters and described in his treatise "*Historia Animalium*". The Greeks used to serve oyster meat in wine and shells as ballot papers. The Athenians also used oyster shells as ballot papers to vote to banish unpopular citizens.

Oyster meat is a food delicacy from time immemorial for the English, Romans Italians and most European countries. According to Guinness World Record for most oysters eaten in three minutes is 187 set by the Norwegian Rune Neri in 2003.

3.2 BIOLOGY

The oyster has a heart with three chambers only. The blood is cloudless indicating absence of hemoglobin. Oyster has a pair of kidneys. Oysters are bisexual or intersexual. Hence, they are called protandrous alternating hermophrodites meaning that they start life cycle as male producing sperms, then they switch to female producing eggs. After egg production they switch back to male again. This process they repeat several times in their life span. Among the genera in the "*Ostreagenus*" eggs produced are held in the gills and mantle cavity. The male discharge sperms in the surrounding waters (laviparous) where the eggs are seen. The sperms are drawn, and eggs are fertilized. The fertilized eggs are held by the oysters and incubated for about 7 to 10 days. They are then expelled to begin their veliger stage in the sea. The genus *Crassostrea* the reproduction is slightly different. The eggs are discharged into the sea water and also the sperms. External fertilization takes place in the sea. So, they are called oviparous. In a spawning season a female oyster can release over I million eggs.

Today oyster farming is an established industry all over the world. The Chinese are the first to raise oysters artificially in ponds and in closed water bodies on a commercial way.

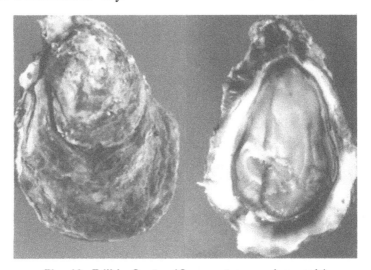

Fig. 10: Edible Oyster (*Crassostrea madrasensis*)

Oysters are bivalve mollusks. They belong to the family of *Oysteredae* and habitats of both marine and brackish waters. There are two varieties of oysters seen in nature. They are classified as:Edible and non-edible.

The edible oysters belong to the following genera: *Oystrea, Crassostrea, Ostrrola* and *Saccostrea.*

Following are some of the important species seen:

- Belon oyster
- Olympia oyster
- Pacific oyster
- Sydney rock oyster.
- Wellfleet oyster and,
- Indian oyster.

The well-known Indian edible oyster is *Crassostrea madrasensis*, seen all over Indian coast line particularly abundantly in the East coast of India.

Pearl oysters

Almost all pearl oysters yield pearls, but only few are used for pearl production by farming. The largest and most popular pearl yielding oyster is the marine species *Pinctada maxima*. The well-known Indian oyster used for farming is the species *Pinctada fucata*.The largest oyster seen in the sea is the *Pinctada maxima*. This grows to the size of a large dinner plate.

The world list of Marine species of pearl oysters includes the following genus:

- *Pinctada albino* (Lamarck, 1819)
- *Pinctada capensis* (Sowerby III, 1980)
- *Pinctada chemnitzii* (Philippi, 1849)
- *Pinctada cumingii* (Reeve, 1857)
- *Pinctada fucata* (Gould 1850) – Akoya pearl oyster- now accepted as *Pinctada imbricate fucata.*
- *Pinctada galtsoffi* (Bartsch, 1931) *Pinctada imbricate* (Roding, 1798) – Gulf pearl oyster.
- *Pinctada inflate* (Schumacher, 1817).
- *Pinctada longisquamosa* (Dunker, 1852)
- *Pinctada maculate* (Gould, 1850)
- *Pinctada margaritifera* (Linnaeus, 1758)- Black-lip oyster
- *Pinctada maxima* (Jameson, 1901)- White-lip or gold- lip oyster
- *Pinctada mazatianica* (Hanley, 1856)
- *Pinctada nigra* (Gould, 1850)
- *Pinctada petersii* (Dunker, 1852)
- *Pinctada radiata* (Leach, 1872)
- *Pinctada sugillata* (Reeve, 1850)
- *Pinctada vidua* (Gould, 1850)

3.3 OYSTER AS FOOD

Oyster is used as a food item by people in coastal areas. There are historical records indicating that oysters have been used by man as a food from pre-historical times. In recent times oyster meat has become a delicacy in view of its nutritional value due to high content of vitamins and many trace elements present in the meat

Methods of consumption

Eat the meat in raw form is still a cherished way of consumption in many parts of the world. Fresh oysters are required for this form of consumption. Oysters have a high shelf life if kept stored alive in ambient waters. But this method of storage needs frequent oxygenation by blowing either oxygen or air in order to keep them alive during transport. If kept stored in chilled waters at temperatures around 5-degree centigrade oysters can remain in good condition up to 15 days without much loss in quality. Oysters are not kept stored und frozen storage as this method tends to reduce its taste by loss many flavors bearing compound by way of drip loss. Raw oysters have different flavors based on variation of species. When eaten in raw form usually lime juice, vinegar, sauces or wine are used to sprinkle over the meat prior to eating to enhance the taste Shell-on cooked, canned, smoked, dried and pickled oysters are also preferred form of eating oysters in different parts of the world.

Health hazards

Oysters are seen often infested with parasites, pathogenic microorganisms and contaminated with heavy toxic metals depending on the conditions of environments where oysters are naturally grown or farmed. There have been several reports in the past indicating human health hazards due to consumption of oysters and other shell fishes caught from polluted waters. Depuration is recommended for 24 to 48 hours in good quality waters and certification by health authorities prior to marketing fresh oysters. There is a great risk in eating fresh oyster meat if history of the product and method of handling and processing are not known. Today oyster meat is globally marketed as a heath food in a powder form. The method of production of oyster powder is given below (Fig. 11):

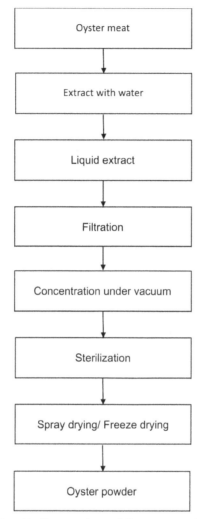

Fig. 11. Processing of Oyster powder

Production of oyster powder

Oyster powder is produced commercially from flesh using advanced biotechnology methods to preserve the active ingredients. Potency of the extract depends largely on the method of extraction. A flow diagram of the method of production of oyster powder is given below:

Oyster powder

In recent times cold extraction using liquid carbon dioxide has become very popular. Using this method very high-quality oyster powder can be obtained with little loss of nutrients.

Table 4: Proximate composition of oyster meat (C. madrasensis)

Composition	g/100 wet tissues (muscle)
Moisture	82.64 ± 1.31
Protein	9.41 ± 0.85
Lipids	2.34 ± 0.32
Carbohydrates	3.20 ± 0.13
Ash	1.01 ± 0.06

3.3.1 Nutritive quality of oyster powder

Oyster powder contains 59 elements, 12 vitamins, 19 amino acids including Taurine, EPA and DHA- awell-known antioxidant glutathione in good quantities. Amino content of oyster meat is given below.

Table 5: Amino content of Oyster Meat

Amino Acids (Crassostrea madrasensis)	g/100g Protein
Aspartic acid	79.33
Threaonine	33.87
Serine	33.73
Glutamic acid	116.00
Proline	35.20
Glycine	43.33
Alanine	47.93
Valine	26.73
Cystine	88.67
Methionine	21.33
Isoleucine	35.40
Leucine	74.30
Tyrosine	12.87
Phenyl alanine	30.73
Histidine	25.20
Lysine	153.93
Arginine	40.73
Tryphtophan	26.13
Total	912.93
Leucine/IsoleucineRatio	2.096
WHO/FAO 1991brecommended ratio	2.36

Taurine is a free amino acid present in mammalian meat and is utilized in biological systems for the elimination of cholesterol from the body through bile acid conjugation. It is also involved in several metabolic path ways and membrane stabilizing properties. It is also a powerful anti-oxidant. Taurine is seen in very high amounts in oyster meat. Some of the pharmaceutical properties attributed to oyster meat may be due to its high taurine content.

Taurine and cholesterol contents of oyster meat.

- Taurine (mg/100g meat): 243 ± 3.5
- Cholesterol (mg/100g meat): 106 ± 2.2
- Taurine Cholesterol Ratio: 2.3

3.4 HEALTH BENEFITS

Several health benefits have been attributed to oyster meat when consumed. But many of these claims are not supported by clinical trials. Some of them are given below:

- Clinically proved that oyster meat increase energy levels in humans and relieves symptoms of fatigue.
- Promotes sexual health in man and women
- Strengthens immune system
- An antidote for common diseases like flu and cold
- Promotes collagen and Keratin synthesis to get healthy hair and nail
- Promotes glutathione synthesis(proved clinically)
- It is an antioxidant
- Corrects erectile dysfunction in man
- Serves as a safeguard against prostate enlargement
- Relieves menstrual pain of women
- Helps liver cleansing
- Promotes secretion of bile
- Improves cardiovascular health
- Helps to cure arthritis and rheumatism
- Normalizes blood pressure

Antioxidant activity

Oyster meat has excellent antioxidant properties. It is attributed to the high content of glutathione in the oyster meat. Glutathionine is an important antioxidant produced in our body. Glutathionine protects cells from oxidative damages. Oxidative damages taking place in the mitochondria are responsible for reduction of energy production in the cells. As a consequence, body shows symptoms of fatigue, decreased immunity, premature aging, less sexual urge etc. Oyster concentrate when given in powder form is reported to relieve all these symptoms and gives an overall wellbeing. Oyster peptide extract is found to have high antioxidant activity as reported by studies conducted by Central Institute of Fisheries Technology, Kochi (Annon, 2012). Dipicryl phenyl hydrazine (DPPH) scavenging assay was carried out in the oyster extract to establish antioxidant assay. IC $_{50}$ value (amount of the extract at which 50% inhibition of DPPH free radical scavenging takes place) for the oyster extract was calculated to be 0.4 mg. This quite high compared to the anti-oxidant activity of gallic acid normally used as standard.

Sample	DPPH IC $_{50}$ value (micro g)
Gallic acid	2.78
Oyster peptide extract	40

Data courtesy, CIFT Kochi

Zinc content

Oyster meat is best source for natural zinc. Oyster meat contains 10 times more zinc than in any red meat of all animals reported so far. Zinc when taken as a metallic constituent externally, as in many multivitamins, causes depletion of copper and other elements like calcium, iron etc. Apart from zinc oyster meat contains several trace elements. Trace elements play an import role in several metabolic activities. These trace elements facilitate transport of several molecules like peptides, amino acids etc. in our body for biosynthesis of body components like protein, lipids etc. When these trace elements are consumed in the form of synthetic supplements they need a binder to transport molecules during digestion and absorption. Hence their biological availability is limited. But minerals present in fish, shellfish, oyster, mussel etc are easily absorbable, being in natural form and beneficial to man.

3.5 OYSTER AND PROSTATE HEALTH

It is now well recognized that poor blood circulation is responsible for prostate enlargement. People with sedentary habits and lethargic in nature get enlarged prostate. Most of them are benign. Enlargement of prostate causes obstruction of bladder and patient experience pain in passing urine. Nitric oxide produced in our body is responsible dilation of blood vessels. Our blood vessels have to remain dilated to facilitate free flow of blood. Oyster meat contains nitric oxide and it supports to retain health of blood vessels to increase better flow of blood.

Menstrual flow

Correct menstrual cycle is a sign of good female health. Oyster extract contains good quantity arginine. Arginine promotes production of nitric oxide in the human body and in turn smoothen muscles and dilate villus veins of uterus. This facilitates and stimulates menstrual discharge and thus improves the condition of obstructed mensus. Hence oyster meat is the best natural food for women to correct their menstrual cycle and associated disorders.

3.5.1 Fatty acids

Oyster meat contains several fatty-acids, particularly the odd numberedsaturated ones. Among the poly unsaturated fatty acids arachidonic acid, C20 :4n6, is seen in very high levels compared to many fish species of marine origin (Table 6).

Table 6: Fatty acids composition of oyster meat

Fatty acid	Mg/100g meat
C12 :0	11.50 ±0.21
C14 :0	4.71 ± 0.04
C15 :0	2.07± 0.29
C16 :0	27.80± 0.03
C17 :0	11.90± 0.22
C18 :0	30.60± 1.40
C20 :0	91.60± 2.60
C21 :0	2.13± 0.01
C23 :0	5.80± 0.05
C14 :1n7	1.42 ± 0.02
C16 :1n7	28.00 ± 0.45

[Table Contd.

Contd. Table]

Fatty acid	Mg/100g meat
C18 :2n6	11.50 ± 0.18
C18 :3n6	4.71± 0.07
C18 :3n3	11.90± 0.18
C20 :2n6	2.07± 0.07
C20 :4n6	27. 80± 0.43
C20 :5n3	112.00± 3.70
C22:6n6	91.60± 2.9
ΣSFA1	188.10 ± 4.85
ΣMUFA2	80.12± 1.50
HPUFA3	261.58± 7.61
Σn-6 FA4	46.80± 0.83
Σn-3FA5	214.90± 6.78
N3/n6	4.66
Total fatty acids	584.20± 13.96

3.6 REFERENCES

1. Ackman, R.G., Eaton, C. A., and Dyerberg, J. (1980). Marine Docosenoic acid isomer distribution in the plasma of Greenland Eskimos. Am. J. Clin. Nutr. 33, 1814- 1817.

2. Andhis-Sobocki, P., Jonsson, B., Wittchen, H.U and Olesseu, J. 2005. Cost of disorders of brain in Europe. European Journal of Neurology, 12(Suppl. 1):1-27.

3. Brenna JT (March 2002). "Efficiency of conversion of alpha-linolenic acid to long chain n-3 fatty acids in man". Current Opinion in Clinical Nutrition and Metabolic Care. 5 (2): 127.

4. Burdge GC, Calder PC (September 2005). "Conversion of alpha-linolenic acid to longer-chain polyunsaturated fatty acids in human adults". Reprod. Nutr. Dev. 45 (5): 581–97.

5. Cangiono, C. *et al.* eating behavior and adherence to dietary prescription in obese subjects treated with 5- hydrxytryptophan . American Journal of Clinical nutrition, 56:863-868, 1992.

6. Den Boer J. A. *et al.* Behavioral neuroendocrine and biochemical effects of 5-hydroxytryptophan administration in panic disorder. Psychiatry Research. 31: 262-278, 1990.

7. Dyerberg, J., Bang., H.O, Hjorne, N.(1975). Fatty acid composition of the plasma lipids in Greenland Eskimos. Am. J. Clin. Nutri. 28, 958—966.

8. Dyerberg, J., Bang., H.O, Hjorne, N.(1997). Plasma cholesterol concentration in Caucasian Danes and Greenland Eskimos. Dan. Med. Bull. 24,52-55.

9. Dyerberg. J, (1982). Observations on populations I Greenland and Denmark. Nutritional Evaluation of Long- Chain Fatty Acids in Fish oil. Edi. S. Barlow anf M. E. Stansby. Academic Press. New York.

10. Elizabeth Mannie. 2006. The truth about Omega-3. Prepared foods. June 2006. Pp 1-7.

11. Genuis S. J. and Schwalfenberg G. K. Time for an oil check up: The role of essential fatty acids in maternal and pediatric health. J, Perinatol. 2006: 26(6) 359-65.

12. Gerster H (1998). "Can adults adequately convert alpha-linolenic acid (18:3n-3) to eicosapentaenoic acid (20:5n-3) and docosahexaenoic acid (22:6n-3)?". Int. J. Vitam. Nutr. Res. 68 (3): 159–73. PMID 9637947.

13. Gopakumar, K. 1997. Biochemical Composition of Indian Food Fishes. Second Edition, CIFT, Cochin 29.

14. Hibbeln, J.R and Davis, J.M. 2005. Considerations regarding neuropsychiatric nutritional requirements for intake of omega-3 highly unsaturated fatty acids. Prostaglandins, Leukotrienes and Essential fatty acids. 81(2) : 179-156.

15. Khan, R.S. *et al.* L-hydroxytryptophan in treatment of anxiety disorder. Journal of Affective Disorders, 8: 197-200, 1995.

16. Kwak SM, Myung SK, Lee YJ, Seo HG (2012-04-09). "Efficacy of Omega-3 Fatty Acid Supplements (Eicosapentaenoic Acid and Docosahexaenoic Acid) in the Secondary Prevention of Cardiovascular Disease: A Meta-analysis of Randomized, Double-blind, Placebo-Controlled Trials". Archives of Internal Medicine. 172 (9): 686–94.

17. Logan AC. (November 2004). Omega-3 fatty acids and major depression :a primer for mental health professional. Lipids in Health and Disease. 3: 25.doi:10.1186/1476-511X-3-25, PMID 15535884.)

18. Lohner, S.; Fekete, K.; Marosvölgyi, T.; Decsi, T. (2013). "Gender differences in the long-chain polyunsaturated fatty acid status: systematic review of 51 publications.". Ann Nutr Metab. 62 (2): 98–112.

19. Lovington. M. B. Omega-3 fatty acids. Am. Farm Physician. 2004.70. 133-40.

20. MacLean CH, Newberry SJ, Mojica WA, Khanna P, Issa AM, Suttorp MJ, Lim YW, Traina SB, Hilton L, Garland R, Morton SC (2006-01-25). "Effects of omega-3 fatty acids on cancer risk: a systematic review.". JAMA: The Journal of the American Medical Association. 295 (4): 403–15.

21. Miller PE, Van Elswyk M, Alexander DD (July 2014). "Long-chain omega-3 fatty acids eicosapentaenoic acid and docosahexaenoic acid and blood pressure: a meta-analysis of randomized controlled trials". American Journal of Hypertension. 27 (7): 885–96.

22. Morris, M.C., Evans, D.A, Tangney, C.C., Bienis, J.L and Wilson, R.S. 2005. Fish Consumption and congenitive decline with age in large community study. Archives of Neurology, 62(12): 1849-1853.

23. Odeleye OE and Watson RR. Health implication of the n-3 fatty acids. Am. J Clin. nutr 1991;53; 177-8.)

24. Peet, M and Stokes, C. 2008. Omega-3 fatty acids in treatment of psychiatric disorders. Drugs. 65(8): 1051-1059.

25. Psychoneuroendocrinology 2011 Apr., 36(3). 393-405. Epub. 2011 Ja. Ev. PMID. 21257271 (Pub Med- index for Medline).

26. Reeves AM, Austin MP and Parker G., April 2005. Role of omega-3 fatty acid as a treatment for depression in perinatal period. The Australian and New Zealand Journal of Psychiatry 39 (4):274-80.

27. Rizos EC, Ntzani EE, Bika E, Kostapanos MS, Elisaf MS (September 2012). "Association Between Omega-3 Fatty Acid Supplementation and Risk of Major Cardiovascular Disease Events A Systematic Review and Meta-analysis". JAMA. 308 (10): 1024–33.

28. Singh. M. Essential Fatty acids, DHA and EPA and human brain. Indian J, Pediatr. 2005.,72(3): 239-42

29. Sourairac, A. *et al.* Action of 5-hydroxytryptphan, serotonin precursor, on insomniacs. Anals Medicon Psychologiques, 135: 792-798, 1997.

30. Van Hiel,L..J. L—5-hydroxytryptophan in depression: the first substitution therapy in psychiatry? The treatment of 99 out patients with therapy- resistant depression. Neuropsychobiology, 6: 230-240, 1990.

31. Waidwer J.Araragi N, Gutknecht L., Lesch KP.: Tryptophan hydroxylase (TPH2) in disorders of cognitive control and emotion regulation a perspective:

32. Young, G and Conquer, J. 2005. Omega-3 fatty acids and neuropsychiatric disorders. Reproduction Nutrition Development. 45(11): 1-28.

33. Young. G. and Conquor. J. Omega-3 fatty acids and neuropsychiatric disorders. Reprod.Nut. Dev. 2005 (1): 1-28.

CHAPTER **4**

OMEGA 3 FATTY ACIDS: THEIR ROLE IN HUMAN NUTRITION

4.1 WHAT ARE OMEGA-3 FATTY ACIDS?

For naming fatty acids in 1953 a new biochemical system of nomenclature was suggested replacing the old Geneva system. According to the new system the unsaturated bonds are numbered with reference to terminal methyl group instead of numbering from the functional group, -COOH. The terminal group is notated as ω (omega) the last Greek letter. The terminal methyl group is numbered 1. The term omega-3 indicates that the first double exists in the 3rd carbon atom with respect to the terminal methyl group (ω)of the long carbon chain unsaturated fatty acid. According to this scheme 4 main families of unsaturated fatty acids are established. For example, Linolenic acid earlier called Alpha- Linolenic acid (ALA), 18:3, is now notated as 18:3 ω 3, meaning there are 18 carbon atoms in the fatty acid, 3 double bonds and the first double bind exists in the 3rd carbon atom with respect to the terminal methyl group(ω).According to this classification, 4 main families of unsaturated fatty acids are established. They are (Fig. 12):

Palmitoleic acid   C16:1W7 $CH_3(CH_2)_4CH_2$ — $CH_2(CH_2)_5CH_2$ — OH

Oleic acid   C18:1 w 9 $CH_3(CH_2)_6CH_2$

Linoleic acid   C18: 2 w 6 HO ... $CH_2(CH_2)_3CH_3$

Linolenic acid C18:3w3 $CH_2(CH_2)_5CH_2$ — OH ... CH_3

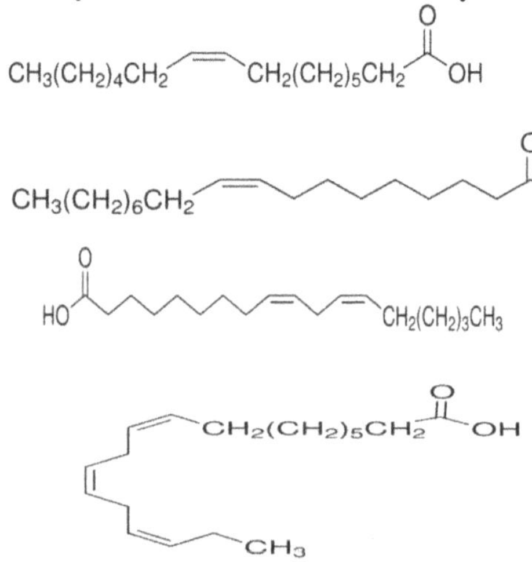

Fig. 12. Families of polyunsaturated fatty acids

Table 7. Unsaturated fatty acid families

ω-3 Unsaturated	ω-6 Unsaturated	ω-7 Unsaturated	ω-9 Unsaturated
α-Linolenic (18:3)	Dihomo-γ-linolenic (20:3)	Vaccenic (18:1)	Gondoic (20:1)
Stearidonic (18:4)	Arachidonic (20:4)	Paullinic (20:1)	Erucic (22:1)
Eicosapentaenoic (20:5)	Adrenic (22:4)	Oleic (18:1)	Nervonic (24:1)
Docosahexaenoic (22:6)	Osbond (22:5)	Elaidic (trans-18:1)	Mead (20:3)
	Palmitoleic (16:1)		
	γ-Linolenic (18:3)		

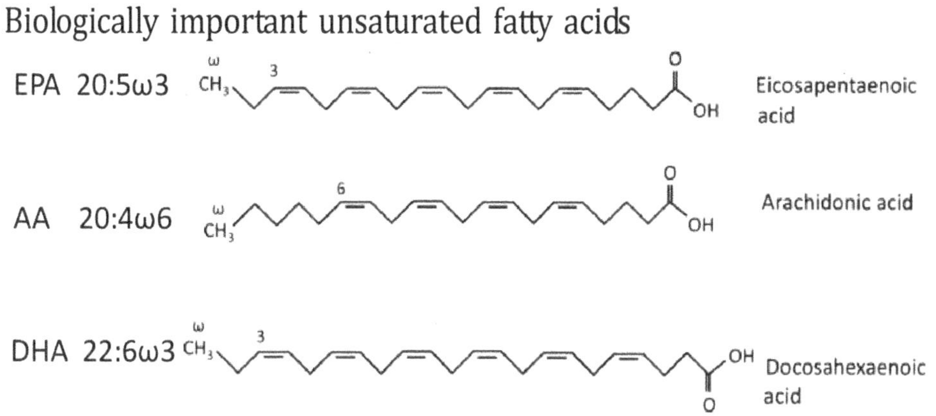

Biologically important unsaturated fatty acids

EPA 20:5ω3   Eicosapentaenoic acid

AA 20:4ω6   Arachidonic acid

DHA 22:6ω3   Docosahexaenoic acid

Fig. 13: Unsaturated fatty acid

There are 11 omega-3 fatty acids identified existing in nature. They are listed below:

Table 8: Lists of the most common omega-3 fatty acids occurring in nature

Nos.	Common name	Chemical	Formula	IUPAC nomenclature
1	Hexadecatrienoic acid (HTA)	16:3 (n-3)		*all-cis*-7,10,13-hexadecatrienoic acid
2	α-Linolenic acid (ALA)	18:3 (n-3)		*all-cis*-9,12,15-octadecatrienoic acid
3	Stearidonic acid (SDA)	18:4 (n-3)		*all-cis*-6,9,12,15-octadecatetraenoic acid
4	Eicosatrienoic acid (ETE)	20:3 (n-3)		*all-cis*-11,14,17-eicosatrienoic acid
5	Eicosatetraenoic acid (ETA)	20:4 (n-3)		*all-cis*-8,11,14,17-eicosatetraenoic acid
6	Eicosapentaenoic acid (EPA)	20:5 (n-3)		*all-cis*-5,8,11,14,17-eicosapentaenoic acid
7	Heneicosapentaenoic acid (HPA)	21:5 (n-3)		*all-cis*-6,9,12,15,18-heneicosapentaenoic acid
8	Docosapentaenoic acid (DPA),	22:5 (n-3)		*all-cis*-7,10,13,16,19-docosapentaenoic acid
9	Docosahexaenoic acid (DHA)	22:6 (n-3)		*all-cis*-4,7,10,13,16,19-docosahexaenoic acid
10	Tetracosapentaenoic acid (TPA)	24:5 (n-3)		*all-cis*-9,12,15,18,21-tetracosapentaenoic acid
11	Tetracosahexaenoic acid (THA)	24:6 (n-3)		*all-cis*-6,9,12,15,18,21-tetracosahexaenoic acid

4.2 WHAT ARE ESSENTIAL FATTY ACIDS?

Fat is an obvious source of energy for animals. In the animal body fats are oxidized or burnt to get energy for all biological needs like movements, synthesis of compound etc. Excess fat is stored as depot fat and animals undergo starvation this stored fat is used to get energy. In this process fatty acids are selectively oxidized by the animal to carbon dioxide, water and energy.

Human body can synthesize most of the fatty acids but not all. Some of the fatty acids which are very essential for the physiological activities and synthesis of some vital components of body are called essential fatty acids. Deficiencies of these acids are responsible for a host of symptoms, diseases, malfunctioning of important organs like liver, kidneys, heart, changes in blood lipid profile, weakening of immune system, depression, dryness and scaliness of the skin, clotting mechanism of blood, inflammation etc. They belong to the group of ù3 and ù6 fatty acids.

Supplementation essential fatty acids can reverse all the above symptoms to a large extent. Clinical studied conducted have already shown the beneficial effects of omega-3 fatty acids in prevention of atherosclerosis, prevention of heart attack, reduction of joint pain in osteoarthritis and some disorders of the nervous system. Health researchers have shown that essential fatty acid deficiency contributes to psychiatric neurological disorders, particularly childhood neurodevelopment disorders including attention Deficient Hyper-activity(ADHD), dyslexia, dyspraxia/developmental coordination disorder and autistic spectrum disorder. Medical applications of these applications are not supported or approved by agencies like US FDA or WHO. But omega-3 enjoys a lucrative market all over the world because effects are conspicuous and satisfactory to users. Most of them are marketed as food supplements or nutraceuticals.

4.2.1 All are omega-3 acids equal in functions and applications?

All omega-3 acids are not created equal. Modern research has revealed that most important fatty acids are the DHA and EPA. All other essential acids like the Alpha- Linolenic acid have to be converted by body to either DHA or EPA to become biologically active and this conversion is very poor. This necessitates that to overcome essential fatty acid deficiency it is important that supplementation should be only DHA and EPA. Two of the important essential fatty acids widely consumed by man come from plant origin. They are alpha-Linolenic acid (ALA, 18:3 ω3) and Linoleic acid (LA, 18:2ω6). These two acids are not produced in human body as our body lacks the enzymes to desaturate other fatty acids to synthesize them. Hence, they have to be supplemented through diet. Most vegetable oils like sunflower oil, olive oil, peanut oil, safflower oil are rich sources of these acids. Alpha-Linolenic acid is rich in vegetables, beans, nuts, seeds, and fruits. Flax seed oil and flax seeds are rich source of ALA (Table 9).

Table 9: Omega-3 acids content of common vegetable oils (%)

Vegetable Oils	Omega 3 Acid Composition
Flax seed	53-63
Linseed oil	53
Canola oil (rape seed oil)	11
Walnut oil	10
Wheat germ oil	7
Soyabean oil	7

Investigations have also shown that omega-3 acids in vegetables, fruits and beans are more in a stable form compared other sources. (Odeleye OE and Watson RR.1991)

Omega-6 fatty acids.

Linoleic acid.

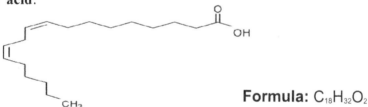

Formula: $C_{18}H_{32}O_2$

Molecular weight: 280.4455

Fig 14: Chemical structure of Linoleic Acid

Linoleic acid (18:2 *poly-unsaturated* fatty acid, due to the presence of two C=C double bonds. It is one of the main fatty acids found in vegetable oils such as olive oil, soybean oil, corn oil and rapeseed oil. The word *linoleic* comes from the Greek word *linon* (flax), and *oleic refers to olive oil in which it is naturally occurring fatty acid*

Leafy vegetables, seeds, nuts, grains and vegetable oils are rich source of Linoleic acid. Sunflower oil, safflower oil,ame oil, palmolein oil, meat, milk and eggs are the richest natural source of this acid. Gamma-linolenic acid (GLA, 18:3), dihomo-gamma-linolenic acid DGLA, 20:3) and arachidonic acid (AA, 20:4) are other important omega-6 fatty acids of nutritional importance to man.

4.3 METABOLISM OF OMEGA-3 UNSATURATED LONG CHAIN FATTY ACIDS

Enzymes present in the human body change the alpha linoleinic acid to EPA and DHA. But this conversion rate is poor. Only 4-5 % of the ALA is converted. During early part of life, human body has only limited capability to convert ALA into the biologically active essential fatty acids DHA and EPA.

4.3.1 Conversion efficiency of ALA to EPA and DHA

Humans can convert short-chain omega-3 fatty acids to long-chain forms (EPA, DHA) with an efficiency below 5%. (Gerster, 1998, Brenna, 2002,) The omega-3 conversion efficiency is greater in women than in men, but less-studied (Burdge, 2005). Higher ALA and DHA values found in plasma phospholipids of women may be due to the higher activity of desaturases, especially that of delta-6-desaturase(Lohner *et al.* 2013). Hence it is now advised that during fetal life by mother and in infancy DAH and EPA should be consumed (Singh. M. 2005, Lonvington.M. B. 2004, Young and Conquer, 2005., Genuis and Schwalfenberg., 2006).

Importance of omega-3 and omega-6 fatty acid ratio

Both Omega-3 and omega-6 groups of fatty acids are metabolized by same enzymes. Hence increase in the amount of one group can decrease the metabolism of another group. There for it is essential to maintain an optimal ratio in the diet to get the normal growth and development and also to help the human body to fight against diseases like heart disease, elevated levels of serum cholesterol and triglycerides. The ideal ratio of omega-6 to omega-3 is 5-10: 1 (Singh. M. 2005).

Imbalanced consumption of EFAs contributes to a wide range of diseases.The role of omega-3 and omega-6 are divergent and contrary. Omega-6 is pro-inflammatory. Omega-3 is anti-inflammatory. A ratio above 50 is harmful to health. Normal diets, particularly for vegetarians, certain high amounts of omega-6 acids compared to omega-3 fatty acids. For western diet the estimated ratio of omega-6 to omega-3 fatty acids are about 20:1. Most Indians eat a diet very rich in omega-6 fatty acids. The ratio of omega-6 to omega-3 in an average Indian diet is reported to be about 30-70:1 (Singh. M. 2005). Indians have a low intake of monounsaturated fatty acids, n-3PUFA and fiber. Most Indian diets have a high content of saturated fats, carbohydrates and trans fatty acids due to high intake of hydrogenated oil. Such a kind of intake of unbalanced fatty acids predisposes Indians to an increased risk of diseases like Coronary Heart Disease (CHD), hypertension, obesity, *diabetes mellitus* and even cancer. The nutrient imbalances result among Indian population higher triglycerides in blood (dyslipidemia) and inflammation. This inflammation makes Indian population vulnerable to Type-2 *diabetes mellitus* and early onset of artherosclerosis. The number of Indians undergoing by-pass surgery is increasing at an alarming rate today. Increased consumption of Omega-3 PUFA, DHA and EPA, can considerable improve the current situation. But consumption of ALA of vegetable original to DHA and EPA by chain length extension and desaturation is very poor in man. Hence, marine fish and fish oil appear to be only source to reduce this nutritional imbalance among Indian population (Table 10).

Table 10: Omega – 3 acids in fish body fat of Indian fishes (*Gopakumar,*1997)

Acids	EPA	DHA
Oil Sardine (*Sardinia longiceps*)	16.2	24.7
Anchvies	3.8	25.2
Pomfret (white)	8	8.5
Cat fish	4.1	15.5
Seer fish	3.2	10.5
Bombay duck	11.2	23.1
Tuna (skip jack)	6.7	30.9
Green mussel (*Perna viridis*)	10.7	15.7
Prawn (*Paeneus indicus*)	14.2	8.2
Squid (*Loligo spp.*)	7.4	31
Milk fish (*Chanos chanos*)	2.1	1.2
Crab	12.4	17
Rock lobster (*Panulirus spp.*)	15.4	15.3

Studies on Greenland Eskimos

There are several documented reports showing that the Eskimos have a tendency for bleeding." The Eskimos are strongly inclined to bleeding. The tendency almost amounts to haemophilia"(Norman- Hansen 1911). But it was Porch, a German Physician, as early as in 1856 first to document that the Eskimos suffer from haemoptyses and this may be due to their dietary habits of consuming food of only animal origin. Peder Freuchen, a well-known Arctic explorer and adventurer also reported that the Greenland Eskimos suffer from frequent nose bleeding and he had pointed out that this may be due to eating foods exclusively of animal origin. Dyerberg *et al.* (1975, 1977, 1980) in their extensive investigation spread among Eskimos population in several districts of Greenland observed that death due to Acute Myocardial Infraction(AMI) and Ischemic Heart Disease are rare compared to Danish population. Their bleeding time also is high. This is reasoned out to be due to their high intake of marine food particularly the blubber rich in marine oil. The Eskimos have also found to have lower triglyceride content and lower levels of cholesterol due to lower LDL and VLDL concentrations. Dyerberg (1982) also observed that the expatriated Greenland Eskimos living in Denmark did not differ in their plasma lipid and lipoprotein levels from the Danes. This investigation has brought to limelight an important fact that plasma lipid profile can be heavily influenced by dietary habits and that prolongation of bleeding time is due to a decreased platelet aggregability paralleled with a shift in platelet fatty acid composition from omega-6 to omega-3 family. The diet of Eskimos is rich in C20: 5w3 fatty acid (EPA), the precursor of the prostaglandin-3 series. The increased synthesis of the prostaglandin of the w-3 series due to higher consumption of w-3 fatty acids rich marine foods is responsible for the decreased thrombotic tendency in Eskimos. These observations brought to our knowledge the importance of omega-3 fatty acids in our diet particularly the DHA and EPA in preventing thrombotic tendency and preventing IHD, lowering of blood cholesterol levels and many other health benefits.

Heart attacks and strokes

Thrombosis, clotting of blood, is the chief cause of heart attacks and strokes. Prostaglandin H2 (PGH2) present in the blood platelets caused dense aggregation of platelets. This is mediated by thromboxanes derived from prostaglandins. If the production of prostaglandin is in excess, more than normally required, it will result in severe and rapid production of blood clots leading to death. In a healthy human normally this does not happen. Our blood vessel walls produce another group of prostaglandins (PGI2) that disaggregates the platelets in an important physiological regulatory process. This agent the PGI2 is produced by the endothelial cells lining of blood vessels. This is a normal process. When this mechanism is reduced or stopped only then strokes takes place and if not immediately controlled it becomes fatal. In a healthy person both these processes are in a balanced equilibrium and we experience neither excess thrombosis or bleeding. Both thrombosis and bleeding tendencies are produced by enzymes, cyclo-oxygenases (COX 1 and 2). There are number compounds available in the market used for therapeutic control of these two diseases. Most of them are enzyme inhibitors. Aspirin is a well-known compound extensively used in clinical practice for prevention of blood clots or thinning blood and prevention of blood clots. Aspirin is a powerful cyclo-oxygenase inhibitor. But prolonged use of aspirin can result in increase of bleeding time.

4.4 BENEFICIAL EFFECTS OF OMEGA-3 ACIDS

Cardiovascular diseases

Extensive clinical investigations carried out on the beneficial effects of omega-3 acids on man and animals have shown that these acids can considerably lower many risks associated with cardiovascular diseases. The first classical work in this field was done by Dyerberg (1982 and Dyerberg et al. (1975, 1977) and Ackman et al. (1980) on Greenland Eskimos. Dyerberg (1982) observed that fish oil as a major factor responsible for the low incidence of thrombotic disorders in Eskimos. He had reported in this study that supplementation of EPA – concentrate resulted in a significant increase in Antithrombin- III concentration from 0.33 to 0.36 g/l and this is due to dietary influence of omega-3 acids. Increase in Antithrombin – III gives resistance against thrombotic disorders among Greenlan Eskimos.

However, many studies conducted by several authors did not support the beneficial role for omega-3 fatty acid supplementation in preventing cardiovascular disease including myocardial infarction and sudden cardiac death or stroke (Rizos et al., 2012; Kwak et al., 2012). Consumption of omega-3 acid was found to lower blood pressure, both systolic and diastolic, in people suffer from hypertension (Miller et al.) Recent epidemiological studies reported and clinical studies conducted (GISSI, 1999, Burr et al., 1989), it is generally accepted that n-3 PUFA have their most useful role in prevention of CVD and also an improvement in the HDL: LDL (Dewailly et al., 2003).

Today formulations of EPA and DHO enjoy a lucrative market all over the world. They are also prescribed by medical practitioners and also freely purchased by consumers from markets without any medical advice. Fish oils of sardines, mackerel, anchovies and salmon, once a low-priced commodity and used in the past for other commercial purposes today are highly priced ones. They undergo enrichment process to enhance concentration of DHA and EPA, purification and encapsulation in gelatin capsules. Price of the product depends on the stage of purification and enrichment of PUFA content. Many consumers are not aware of this aspect. Some of the important cardiovascular diseases for which omega-3 acids are used as therapeutic agent are listed below. But none of these applications are approved by certifying agencies like WHO, US FDA or other national agencies. So far only one application met the approval of USFDA for lowering of serum triglycerides for Ethyl esters of EHA and DPA for Glaxo-Smith – Kline under prescription of clinical practitioners in 2010 in USA.

4.5 APPLICATIONS

- Prevention of arrhythmias that lead to sudden cardiac death.
- Prevention of thrombosis (blood clot formation)
- Lowering of serum triglycerides
- Reducing risk of atherosclerosis (plaque formation in arteries due to deposition of cholesterol esters of saturated fatty acids)
- For general improvement of Vascular endothelial function
- Reduction of inflammation
- Lowering of blood pleasure
- Dermatological applications to reduce skin deceases due to essential fatty acids deficiency.
- To improve bleeding time in human system by enhancing synthesis of prostacyclin's.

Recently World Health Organization and a number of agencies have identified that deficiency of omega-3 fatty acids, EPA and DHA, is a serious issue and these acids are required in adequate quantities in human diet to maintain for the health of heart, brain, normal growth and development (Elizabeth Mannie, 2007). The American Heart Association (AHA) recommends that adults who solely consume plant-based omega-3 fatty acids alone should consume fish twice a week. If they are vegetarians, they have to consume omega-3 capsules to get the required nutrition. There are now capsules made of vegetable gelatin also. This due to the finding that plant-based omega-3 acids are of the lower molecular weight ones like the Linoleinic (C18:3 w3) and they have to be converted to EPA and DHA to be biologically active for human being and this conversion is only very poor in man. The AHA in their 2003 recommendation indicated that 2-4g of EPA and DHA daily can lower serum triglycerides by 20-40%. This effect appears to be synergistic with the HMG-CoA reductase inhibitor(statin)drugs such as simvastin (Zocor), pravastatin (Pravachol) and atorvastatin (Elizabeth, 2006). The AHA also recommended in2003 that people with known coronary disease should take daily approximately 1g of EPA and DHA combined by eating fish or taking fish oil supplements. The risk of cardiovascular diseases increases with age because the arteries become stiffer. Two of the well-known methods commonly used to measure arterial stiffness are Pulse Wave Velocity (PWV) and Augmentation Index (AI). High PWV and AI values indicate that the arteries are stiffer. Modern research in this field has shown that supplementation of omega-3 fatty acid in diets reduces considerably both PWV and AI values people of age group between 60 to 80 years group.

4.5.1 Asthma

It is now shown that leukotrienes play an important role in the pathology of asthma. Leukotrienes are formed from Arachidonic acid (C20:4w6), an omega-6 fatty acid by the enzyme cyclooxygenase-2(Cox2) in the human body. A study conducted on Australian school children (Elizabeth, 2006) showed that consumption of one fish meal per week reduced incidence of asthma compare control group that rarely ate fish. Omega- 6 acids are also important in diet. Today the ratio of omega − 3 to omega − 6 has become critical while diets are formulated. In order to get fuller benefits of omega 3- acids it vital that the presence of omega − 3 should there in the diet because of the specific role of Arachidonic acid in the synthesis of prostaglandins and thromboxanes. In an important clinical study conducted by Mihrashahi *et al.* in 2004 on group of 616 women at the risk of having children with asthma showed that mothers who received fish oil concentrates and give fish oil concentrates to their infants after birth have little nocturnal cough and wheezing compared to control groups who were not given any fish oil. But it should be noted that a number of clinical studies conducted in this area gave only inconsistent results and hence no conclusive support can be given to this view.

4.5.2 Essential fatty acids and their role in mental health

"There is a critical role of EFAs and their metabolic products for maintenance of structural and functional integrity of central nervous system and retina. Most of the brain growth is completed by 5–6 years of age. At birth brain weight is 70% of an adult, 15% brain growth occurs during infancy and remaining brain growth is completed during preschool years. DHA is the predominant structural fatty acid in the central nervous system and retina and its availability is crucial for brain development" (Meharban Singh, 2005).

Mental illnesses increasing globally and according to many experts, they will become a major problem in the developed world. There have been several natural and supernatural reasons assigned to this disease which plague millions of people all over the world. The absence or diminution of light is potentially an anxiety arousing situation for most of us. Even a cursory look at the origin of myths many centuries reveals that chaotic, destructive and dangerous aspects of darkness (Wilson, 2012). There are at least fifty well identified disorders which fit under this disorder called mental diseases. In 2004, mental health diseases took over heart disease as leading disease in Europe and estimated to cost Euro 386 billion (Wittchen and Olesen, 2005).

Some of the common symptoms associated with this disorder are given below:

● Pessimism and feeling of helplessness

● Irritability and intolerance

● Loss of memory

● Difficulty in decision making

● Insomnia and tiredness

● Tendency to over eat resulting in obesity and depression

● Physical problems such as persistent headaches and other pains

● Lethargy and getting things delayed.

Recent studies suggest that consumption of seafood rich in long chain Omega-3 polyunsaturated fatty acids, particularly DHA and EPA, may have positive impact on mental health diseases including Dementia and Alzheimer's disease (Morris *et al.*, 2005, Peet and Stokes, 2005). For the brain to function efficiently, it needs several bio-chemicals. Some of them are produced by our body and some of them are to be supplied externally through diet. With aging, human body's efficiency to produce them also declines. Some of the essential ones are to be supplied through diet. Among them tryptophan, precursor of serotonin, is the principal molecule. Omega-3 fatty acids and a non-essential amino acid L- tyrosine are the other two compounds which play vital role in prevention of mental disorders.

One principal compound responsible or regulation of mood and depression is serotonin. It is an amino acid (5-hydroxytryptamine, 5 HT) produced by our body from an essential amino acid called tryptophan. If there is no adequate supply of tryptophan in our diet production of serotonin is affected. In human body most of the serotonin is present in the intestine. Serotonin is responsible for the intestinal motility, stimulation of nerve endings and neural transmission. If the level of serotonin is too high in our body it has adverse impact on vascular system. It often causes migranes. On the contrary low level of serotonin impairs neurological functions leading to depression.

5-hydroxytryptamine is naturally produced in our body from endogenous tryptophan by a specific enzyme called tryptophan hydroxylase (TPH). This enzyme TPH converts tryptophan into an intermediate 5-hydroxytryptopahan and finally serotonin. This intermediate, 5-hydroxytryptophan, is also biologically very essential as it is found now important for relieving several depressive disorders like reactive depression, panic disorder, insomnia, anxiety and reduction of appetite for people suffering from obesity (Van Heil *et al.*, 1980; Den Boer *et al.*, 1990; Sourlairac *et al.*, 1997; Kahn *et al.*, 1995; Cangiono *et al.*, 1992)

Tryptophan hydroxylase (TPH) is a rate limiting enzyme in 5-HP biosynthesis. Very recently a new TPH is identified called TPH2 providing new insights into the long-recognized but yet not fully understood role of 5-HT stress interaction. Serotonin (5HT) modulates the stress response by interacting with the hormonal hypothalamic- pituitary-adrenal (HPA) axis and neuronal sympathetic nervous system (CNS), Chen and Muller (2012). Based on genetic variation, there is now accumulating evidence that altered function of tryptophan hydroxylase -2 (TPH2), the enzyme critical for synthesis of serotonin (5-HT) in brain, plays a role in anxiety, aggression and depression related personality traits and in the pathogenesis in disorders featuring deficits in cognitive control and emotion regulation (Waider J, Araragi, N, Gutknecht L and Lesch K P, 2011).

Among the omega-3 fatty acids, DHA (docosa hexanoic acid) is vital to maintain the flexibility and fluidity of brain membrane. This property of the membrane is an important characteristic of brain as it enhances the ability of the cells to transmit signals by flexing and contracting. Hence, DHA is the most important fatty acid used by our brain to function properly. Hence, it is very important that we should consume marine fish rich in these Omega-3 polyunsaturated fatty acids. Those who do not consume fish, it is advisable to consume fish oil capsules regularly along with diet.

4.5.3 Alzheimer's disease and Dementia

Dementia is a group of symptoms caused by gradual death of brain cells. Dementia according to Oxford Dictionary is a chronic or persistent disorder of the mental processes caused by brain disease or injury and marked by memory disorders, personality changes, and impaired reasoning. It has its origin in the late 18th century from the Latin words demens, dement- 'out of one's mind'. It is usually a progressive condition (such as Alzheimer's disease) marked by the development of multiple cognitive deficits (such as memory impairment, aphasia, and the inability to plan and initiate complex behavior ("Dementia." *Merriam-Webster.com.* Merriam-Webster, n.d. Web. 26 Apr. 2018). Dementia is diagnosed only when both memory and another cognitive function are each affected severely enough to interfere with a person's ability to carry out routine daily activities. (*The Journal of the American Medical Association*).

This disease is commonly seen in older people. Alzheimer's disease is caused by the formation of amyloid plaque in the brain. Degeneration of nerve cells is also occurring along with plaque formation. Major symptoms seen in Alzheimers patients are loss of memory and confusion. DHA is the fatty acid in the brain, particularly in the shingomylein, a constituent of phospholipids in the brain cells. It is now found that DHA gives protection against Alzheimer's disease. The famous Framingham heart study (Cohort study) supports the view that daily intake of DHA in diet can considerably reduce the risk of the occurrence of Alzheimer's symptoms considerably.

While the overwhelming number of people with dementia are elderly, dementia is not an inevitable part of aging; instead, dementia is caused by specific brain diseases. Alzheimer's disease (AD) is the most common cause, followed by vascular or multi-infarct dementia. Heprevalence of dementia roughly doubles for every five years of age beginning at age 60. Dementia affects about 1% of people between ages 60 and 64,5-8% of all people between ages 65 and 74, up to 20% of those between 75 and 84, and between 30% and 50% of those age 85 and older. About 60% of nursing home patients have dementia (Farlex Partner Medical Dictionary. 2018). Several studies have shown that marine fish eating has some protective action in controlling Alzheimer's disease. Elderly people who ate fish at least once a week had a 60% lower risk of developing the disease over a 4 year period (Morris *et al.*, 2003). Conquer *et al.* (2000) found lower levels of plasma phospholipid DHA in patients with Alzheimer's disease and other dementias. The authors hypothesized that decreased ω 3 PUFA prior to disease onset may be at least partly responsible for the lower levels of plasma DHA and the cognitive decline. But the potential usefulness of DHA as a drug for treatment of dementia has to be validated by more intensive clinical studies.

4.5.4 Depression and Bipolar Disorder

Unipolar depression and bipolar disorder are recognized as distinct psychiatric condition. It is reported (Report of the Linus Pauling Institute, 2010) that ecological studies conducted across different countries suggest an inverse association between seafood consumption and national rates of people having depression and bipolar disorder. Low concentrations of DHA in cerebrospinal fluid result in lowering of levels 5-hydroxy indole acetic acid. It is reported that low concentration of 5-

HIAA is associated with depression and suicide It is also seen that people who suffer from Schizophrenia have decreased levels of omega-3 fatty acids in their blood cells and brain. Several clinical studies conducted recently indicate that when omega-3 fatty acids supplementation is given to Schizophrenia patients along with drugs normally given there were marked improvements in their condition compared to control groups. Studies on the effect of omega-3 fatty acids, particularly DHA and its effect on the treatment of depression and mental disorders have been a subject of investigation by several workers (Hibbeln 1998, Tanskanen *et al.,* 2001; Sivers and Scott 2002.). Significant correlation was found to exist between high consumption of marine fish consumption and low prevalence of major depression according to the above studies. However, more studies are needed to arrive at definite and conclusive evidences to support the beneficial role of omega-3 fatty acids for its therapeutic effect on these diseases.

4.5.5 Rheumatoid Arthritis

Several studies conducted over the years show that considerable improvements in morning stiffness and join tenderness with regular intake of fish oil supplementation.

4.5.6 Infants and Pregnant women (Post-natal depression)

Many women suffer from postnatal depression. This disease has been found to be due to low intake of DHA in the diet.Postnatal depression has also been linked with a low n-3 PUFA status, which fits with the evidence for high fetal requirements for DHA in the third trimester. Hibbeln (2002) suggests that mothers selectively transfer DHA to their fetuses to support optimal neurological development leaving themselves at risk of depletion if dietary n-3 PUFA intake is low. In a review of 23 countries, Hibbeln (2002) found that both a lower DHA content in breast milk and lower seafood consumption in clinical studies. Llorente *et al.* (2003) supplemented mothers for 4 months after delivery with 0.2 g dayDHA. But he could not find significant effect on self rated depression despite increases in plasma DHA status. Hence this observation need further investigations to establish the role of DHA in mental depression. Human milk is a rich source of DHA and EPA in the ratio 4: 1. Early dietary supplementation of diet rich in DHA and EPA showed improvement on the mental Development Index of children.

4.5.7 Developmental Coordination Disorder (DCD)

This is a disease characterized by the disturbance of perception attention and behavior pattern in young children. DCD was found to affect 5% of school-aged children. The major characteristics of the disease are difficulties in learning, behavior and deficiency in motor function. There are several reports showing that adequate intake of DHA and EPA can bring significant improvement in patients.

4.6 RECOMMENDATIONS OF INTERNATIONAL AGENCIES ON THE LEVELS OF INTAKE OF DHA AND EPA

USFDA has rated that intakes of upto 3g per day of long chain omega-3 fatty acids (DHA and EPA) are generally Recognized As Safe (GRAS) for inclusion in diet and available evidences suggest that intake less than 3g/day are unlikely to result in clinically significant bleeding.

- European Commission (EC)
- EC recommends an omega-6 fatty acid intake of 4-8
- WHO recommends an omega-6 fatty acid intake of 5-8% of energy and DHA and EPA 1-2%.
- Japan Society for Lipid Nutrition(JSLN)
- JSLN recommends that ALA intake may be reduced to 3-4% of energy in Japanese people whose omega-3 fatty acid intakes average 2g/day including 1g of DHA and EPA
- American Heart Association (AHA) recommends that people with documented coronary heart disease eat a variety of fish twice weekly.

4.6.1 Omega-3 and omega-6 fatty acids and their role in synthesis of eicosenoids

Prostaglandins and related compounds are called eicosanoids as they are produced from eicosenic fatty acids. The term eicosenoid is derived from "eicosa" meaning "twenty", referring to the 20 carbons in most of the molecules. Prostaglandin was first isolated from seminal fluid in 1935 by the Swedish physiologistUlf von Euler(19350 and independently by M.W. Goldblatt (1935). The first total syntheses of prostaglandin $F_{2á}$ and prostaglandin E_2 were reported by E. J. Corey in 1969,an achievement for which he was awarded the Japan Prize in 1989.It was later shown that that aspirin-like drugs could inhibit the synthesis of prostaglandins. The biochemists Sune K. Bergström, Bengt I. Samuelsson and John R. Vane jointly received the 1982 Nobel Prize in Physiology or Medicine for their research on prostaglandins.

Most of the prostaglandins are synthesized from arachidonic acid (20:4 Δ5,8,11,14) an omega-6 fatty acid and the triene fatty acid 20:3 Δ8,11,14. Prosataglandins produced from arachidonic acid are called series 2 and the triene fatty acid (20:3) series 1. Prostaglandins derived from series 1 have one double bond less than series 2. Eicosa pentaenoic acid, 20:5 Δ5,8,11,14,17, a fatty acid seen abundantly in marine fish oil is also a substrate for prostaglandin synthesis and the products from this compound have one more double bond than the series 2 products. But the 20:3 Δ5,8,11 cannot be used to make functional prostaglandins.

The eicosanoids are considered local hormones because they have specific effects on target cells close to their site of formation. They are rapidly degraded and are not transported to distal sites within the body.Eicasenoids participate inintercellular signaling. They have roles in inflammation, regulation of blood pressure, blood clotting, modulation of immune system, control of reproductive processes, tissue growth and regulation of sleep/wake cycle.

Examples of eicosanoids

- prostaglandins
- prostacyclins
- thromboxanes
- leukotrienes
- epoxyeicosatrienoic acids.

4.6.2 Role of Prostacyclins

Thromboxane (Tx) A_2 is a vasoconstrictor and platelet agonist. Prostacyclin (PGI_2) is a vasodilator that inhibits platelet function (Yan Cheng *et al.*, 2002.). Aspirin gives cardioprotection through inhibition of TxA_2 formation by platelet cyclooxygenase (COX-1). Prostacyclin (PGI2) generated by the vascular wall is a potent vasodilator, and the most potent endogenous inhibitor of platelet aggregation so far discovered. Prostacyclin inhibits platelet aggregation by increasing cyclic AMP levels. Prostacyclin is a circulating hormone continually released by the lungs into the arterial circulation. Circulating platelets are, therefore, subjected constantly to prostacyclin stimulation and it is via this mechanism that platelet aggregability in vivo is controlled (Moncade S., and Vane JR,1979).

The prostacyclin: thromboxane A2 ratio is important in the control of thrombus formation. By giving small amounts of aspirin this ratio can be altered to effect inhibition of thromboxane formation. The dietary use of a fatty acid like eicosapentaenoic acid (which would be the precursor for a delta17-prostacyclin (PGI3) transforms the platelets into nonaggregating thromboxane A3. Eicosapenaenoic acid is present in rich quantities in marine fish oils like sardine, anchovy and mackerel oils. The antithrombotic properties of fish oil are now attributed to this mechanism inhibition thrombosis. Prostacyclin has interesting potential for clinical application in conditions where enhanced platelet aggregation is involved or to increase biocompatibility of extracorporeal circulation systems.

Role of prostaglandins in Menstruation

The changes of concentrations of prostaglandins (PG) are cyclic in the uterine tissues and related to steroid ovarian hormones. The role in normal menstruation is presumably related to a local haemodynamic effect. PGF2 alpha vasoconstricts the endometrial vessels during menstruation and contracts the smooth muscle of the myometrium. PGE2 vasodilates the vessels of the endometrium, and PGI2 relaxes smooth muscle, vasodilates the vessels of the myometrium and inhibits thrombocyte aggregation (Jensen *et al.*, 1987). During *menstrual* period the uterus contracts to help expel its lining *prostaglandins* are involved in pain and inflammation and they trigger the uterine muscle contractions. Higher levels of *prostaglandins* are associated with more-severe *menstrual* cramps and pains.

4.7 BIOSYNTHESIS OF PROSTAGLANDINS, AND THROMBOXANES FROM ARACHIDONIC ACID

Arachidonic acid (C20:4ω6) is an essential fatty acid seen widely distributed in most mammals. Arachidonic acid is obtained from food such as poultry, animal organs and meat, fish, seafood, and eggs (Taber *et al.*, 1998; Komprada *et al.*, 2005). It is an important component of brain membrane phospholipids.Platelets, mononuclear cells, neutrophils, liver, brain and muscle have up to 25% phospholipid fatty acids as ARA (Calder 2007).Industrially ARA is produced from the filamentous fungus, *Mortierella*, especially of the species *alpine (Zhu et al., 2002; You et al., 2011; Ji et al., 2014)* It performs several important biological functions which include the following;

- Signalling, inflammation and nervous functions.
- Modulation of activities of protein kinases and ion channels
- Inhibition of neurotransmitter uptake and
- Enhancement of synoptic transmission induction of cell adhesion

It has been reported that arachidinic acid is a critical mediator in amyloid – beta induced pathogenesis leading to learning memory and behavioral impairment in Alzheimer's disease ((Sanchez and Mucke 1995). An imbalance in the omega-6 and omega- 3 fatty acid ratio is attributed to be the main reason for inflammatory and auto-immune diseases. Eicosanoids are found to be proinflammatory and immunoregulatory(Bates 1995). Since both eicosanoids and prostaglandins are produced from arachidonic acid, it has profound influence on regulation ofinflammation. The profile of arachidonic biosynthesis was unique to each cell. Endothelial cells synthesized primarily prostacyclin; the lymphocytes synthesized principally thromboxane (Lawrence Levine and Iftekhar Alam. 1979). Aachidonic acid released from phospholipids undergoes two metabolic pathways in leukocytes. The first reaction is catalysis by prostaglandincyclo-oxygenase yielding the prostaglandin endoperoxides GT2 and H2 and thromboxane A2. They induce rapid irreversible aggregation of human platelets and potent inductors of smooth muscle contraction (Malmsten 1886). The second pathway, catalysis by lipoxygenase, yields various hydroperoxy acids. The arachidonic acid is also converted by the enzyme 5- lipoxygenase to leukotrienes. All these reactions put together are called arachidonic acid cascade. Prostaglandin E2 and prostacyclin are potent vasodialators whereas Lekotriene D4 causes cellular adhesion, chemotaxis of neutrophils and degranulation (Samuelsson, 1987).

Arachidonic acid and anti-tumor activity

Undurti Das and co- investigators in a series of studies (1990, 1991, 1995, 1998) have reported that ARA as a potential anti-cancer drug . Thus, ARA was reported to kill tumor cells selectively *in vitro* via eliciting cell surface membrane lipid peroxidation, which can be blocked by vitamin E, uric acid, glutathione peroxidase and superoxide dismutase. ARA is found to suppress proliferation of normal and tumor cells by a variety of mechanisms that may partly depend on the type(s) of cell(s) being tested and the way ARA is handled by the cells.

According to HatemTallima to Rashika El Ridi (2017) ARA levels in pregnant women, infants, children and the elderly in poor rural settings as dietary ARA is safe, being a poor substrate for beta-oxidation and is critically essential for the development and optimal performance of the nervous system, especially the brain and cognitive functions, the skeletal muscle and immune systems. Additionally, ARA promotes and regulates type 2 immune responses against intestinal and blood flukes and may well represent an invaluable endoschistsomicide and endotumoricide.

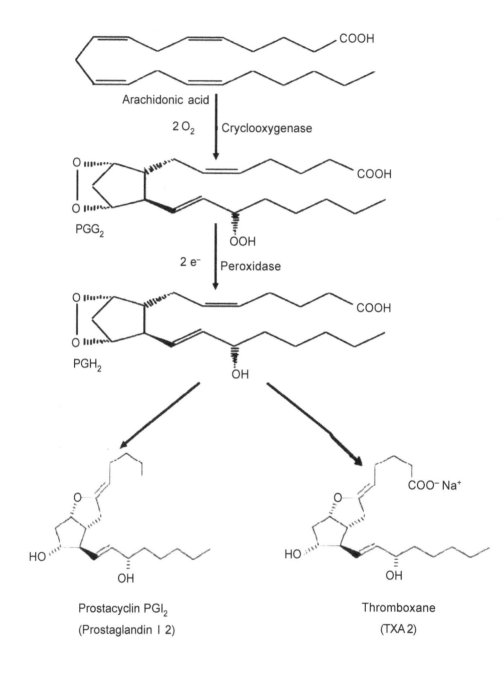

Fig 15. Biosynthesis of Prostaglandins

Three enzymes are involved in the biosynthesis of prostaglandins. They are Phospholipase A2, cyclooxygenase and lipoxygenase. The enzyme cyclooxygenase is abbreviated as COX. Release of the fatty acid from the phospholipid, phosphatidyl inositol, is the first control point in the prostaglandin biosynthetic pathway. Two enzymes are primarily involved in eicosanoid biosynthesis. Prostaglandin synthase and 5-lipoxygenase. Prostaglandin synthase is a complex enzyme that catalyzes the first two steps in the prostaglandin synthesis pathway. It is often called cyclooxygenase (referring to the first of the two reactions it mediates). Three enzymes have been characterized so far. They are COX-1, COX-2, and COX-3.

Humans, and most other mammals have two genes for cyclooxygenase. The products of the genes, COX-1 and COX-2, are structurally quite similar, with only subtle differences. They catalyze the same reactions. Although COX-2 works with a wider range of substrates. COX-1 is constitutively expressed in nearly all tissues. In contrast, COX-2 is inducible, especially by inflammatory stimuli.

Some evidence suggests that COX-1 is responsible for generating the prostaglandins required for protection of the gastrointestinal tract, while COX-2 is responsible for the increased prostaglandin synthesis associated with inflammation, fever, and pain responses. This has led to attempts to find specific inhibitors of COX-2. On the other hand, some evidence suggests that the roles of the two isozymes may not be quite that clearly defined.

A new isozyme, COX-3 was discovered in 2002. It is thought to be an intron-splice variant of COX-1. It has a similar sequence, but not identical amino acid sequence to that of COX-1, but has some functional differences. The role of COX-3 is the subject of considerable interest, but much remains to be investigated.

The neurological deficit and the volume of brain infarction in the group treated with fish oil was less than that of the control group. The present findings suggest that moderate dietary supplements of fish oil may be beneficial in the prophylactic treatment of ischemic cerebral vascular disease Keith *et al.* 1979)Prostacyclin is synthesized from arachidonic acid by the vessel wall of all mammalian species including man. Prostacyclin is a potent vasodilator and the most potent inhibitor of platelet aggregation so far described (Monaca and Vane 1980). It inhibits platelet aggregation through stimulation of an enzyme called adenylate cyclase leading to an increase in levels of cyclic AMP in the platelets. The enzyme which synthesizes prostacyclin is mainly localized in the endothelial layer of the vascular wall. Prostacyclin can also be a circulating hormone constantly released by the pulmonary circulation. On the basis of these observations the above authors propose that platelet aggregability in vivo is controlled via a prostacyclin mechanism. In contrast to the vessel wall, in blood platelets arachidonic acid is converted by the enzyme thromboxane synthetase to a potent vasoconstrictor and proaggregating substance, thromboxane A2. Therefore, arachidonic acid is metabolized in the vessel wall and the platelets to potent substances with opposing biological activities. The balance between the activities of these substances is important in the homeostatic interaction of the platelets and the vessel wall. The balance between thromboxane A2 and prostacyclin might be important in the control of the pulmonary circulation (Moncada and Vane 1980).

Leucotrienes

Leucotrienes (LTS) is a family of lipid mediators. They are classified into two major classes, LTB (4) and cysteinyl leucotrienes (Cyst LTs). They are synthesized in the leucocytes from the parent fatty acid arachidonic acid (C 20:4 w6) by the enzyme 5-lipoxygenase.LTB (4) is one of the most potent chemoattractant mediators of inflammation. It exerts its actions through seven transmembrane-spaning G protein receptors, LTB4 R-1 and LTB4 R-2. CysLTs (LTC(4), LTD(4), and LTE(4)) are potent bronchoconstrictors that play an important role in asthma. They induce their actions through G protein coupled receptors, CysLT R-1 and CysLT R-2. LTs are involved in the pathogenesis of inflammatory disorders specially asthma, rheumatoid arthritis (RA) and inflammatory bowel disease (IBD). Therefore, LTs modifiers, LTs inhibitors or antagonists, represent important therapeutic advance in the management of inflammatory diseases. Zileuton, zafirlukast and montelukast are LTs modifiers that are approved to use for the treatment of inflammatory disorders (Sharma JN and Mohammed LA. 2006). Leukotrienes are released in inflammatory and hypersensitivity reactions Leukotrienes are metabolites of arachidonic acid produced by lungs and leucocytes. Their production is catalysed by an enzyme called 5-lipoxygenase. The first compound produced in the pathway is LTA4. It undergoes two conversions (Sirois and Borgeat, 1984).

• Enzymatic hydrolysis to Leukotriene B4.

• Leukotriene C4 by addition of glutathione by the enzyme glutathione 5-transferase.

Leukotrines C4, D4 and E4 are potent bronchoconstrictors both *in vitro* and *in vivo*. They induce the production of mucus by respiratory tract and decrease its transport. Leukotriene B4 stimulates several leukocyte functions related to

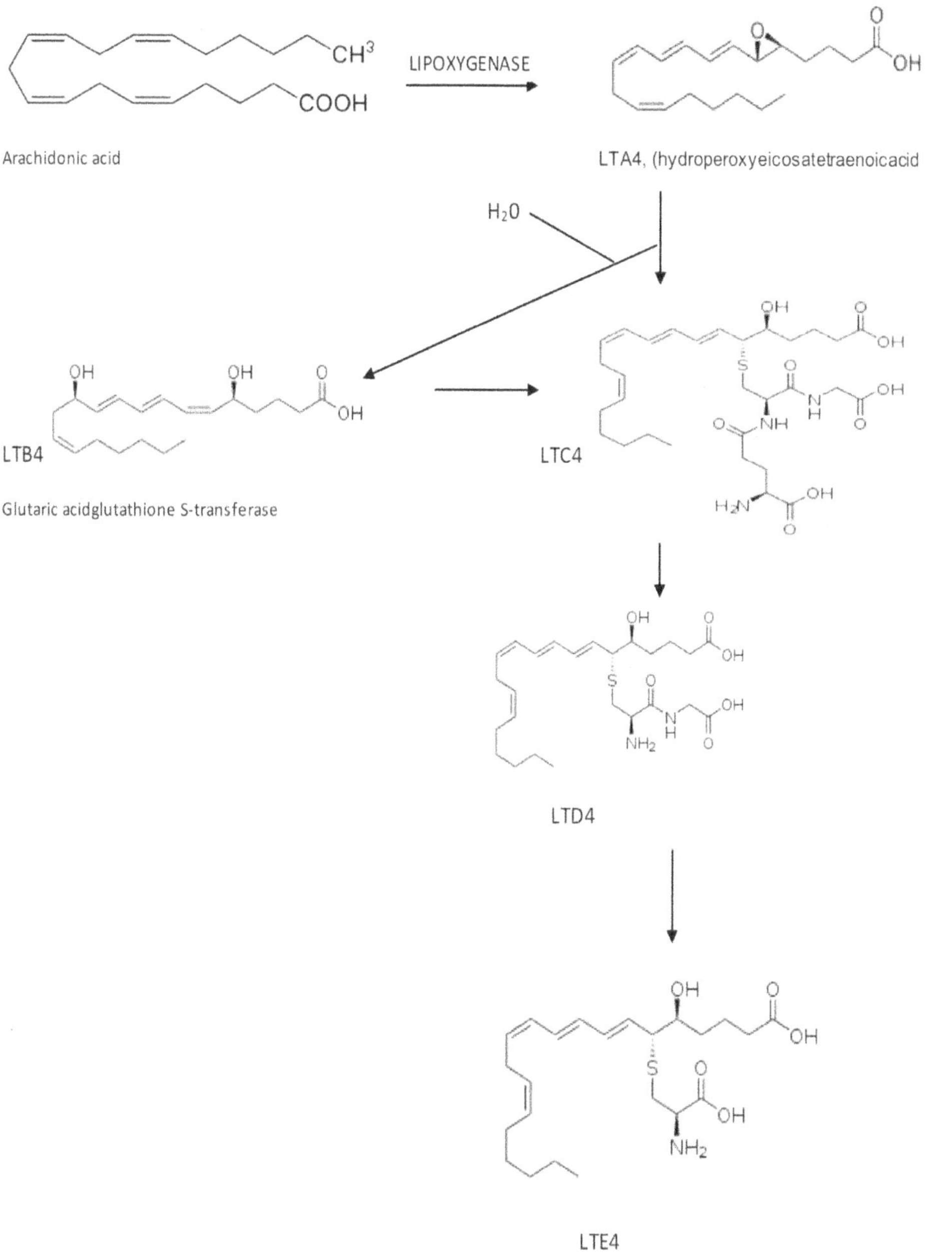

LIPOXYGENASE

Arachidonic acid

LTA4, (hydroperoxyeicosatetraenoicacid

H_2O

LTB4

LTC4

Glutaric acidglutathione S-transferase

LTD4

LTE4

Fig. 16. Biosynthesis of Leucotrienes

inflammation like chemotaxis, aggregation, release of isosomal enzymes and production of superperxide anion and laso induces formation of suppressive and cytotoxic T- lymphocytes (Sirois and Borgeat, 1984).

The cysteinylleukotrienes have long been suspected to play a role in the pathogenesis of asthma. This speculation was based largely on their release in human lung following antigen challenge as well as their potent bronchoconstrictor activity. However, there is increasing evidence that the cysteine leukotrienes also produce several pro-inflammatory effects and alter the activity of neuronal pathways in the airways (Theodore J.T *et al.*, 1995). Cysteinyl leukotrienes stimulatespro-inflammatory activities such as endothelial cell adherence and chemokine production by mast cells. They induce asthma and other inflammatory disorders, thereby reducing the airflow to the alveoli. Acute asthma attacks are often triggered by allergens or exercise. Inflammatory molecules called leukotrienes are one of several substances which are released by mast cells during an asthma attack, and it is leukotrienes which are primarily responsible for the broncho constriction (Drazen, 2003).

Leucotrienes and inflammation

Leukotrienes appear to be the major criteria as mediators of inflammation. They have been shown to be present at a variety of inflammatory sites and to be generated by cells involved in inflammatory sequelae following an inflammatory stimulus. Their effects on the microvasculature and inflammatory cells are consistent with a pro-inflammatory role and the effects of LTB_4 in particular on a range of immune responses may indicate a role in the more chronic phases of inflammatory disease (M. A. Bray 1986).

4.8 REFERENCES

1. Ackman, R.G., Eaton, C. A., and Dyerberg, J. (1980). Marine Docosenoic acid isomer distribution in the plasma of Greenland Eskimos. Am. J. Clin. Nutr. 33, 1814- 1817.

2. Andhis-Sobocki, P., Jonsson, B., Wittchen, H.U and Olesseu, J. 2005. Cost of disorders of brain in Europe. European Journal of Neurology, 12(Suppl. 1): 1-27.

3. Bray, M.A. Agents and Actions (1986) 19: 87. https://doi.org/10.1007/BF01977263.

4. Brenna JT (March 2002). "Efficiency of conversion of alpha-linolenic acid to long chain n-3 fatty acids in man". Current Opinion in Clinical Nutrition and Metabolic Care. 5 (2): 127.

5. Burdge GC, Calder PC (September 2005). "Conversion of alpha-linolenic acid to longer-chain polyunsaturated fatty acids in human adults". Reprod. Nutr. Dev. 45 (5): 581–97.

6. Burr, M.L., Ashfield Watt, P.A., Dunstan, F.D., Fehily, A.M., Breay, P., Ashton, T., Zotos, P.C., Haboubi, N.A. & Elwood, P.C. (2003) Lack of benefit of dietary advice to men with angina: results of a controlled trial. Eur. J. Clin. Nutr. 57, 193–200. Crossref CAS PubMed Web of Science®Google Scholar.

7. Cangiono, C. *et al.* eating behavior and adherence to dietary prescription in obese subjects treated with 5- hydrxytryptophan. American Journal of Clinical nutrition, 56:863-868, 1992.

8. Den Boer J. A. and Westenberg H. G Behavioral neuroendocrine and biochemical effects of 5-hydroxytryptophan administration in panic disorder. Psychiatry Research. 31: 262-278, 1990.

9. Drazen J.M. (2003) Leukotrienes in Asthma. In: Yazici Z., Folco G.C., Drazen J.M., Nigam S., Shimizu T. (eds) Advances in Prostaglandin, Leukotriene, and other Bioactive Lipid Research. Advances in Experimental Medicine and Biology, vol 525. Springer, Boston, MA.

10. Dyerberg, J., Bang., H.O, Hjorne, N. (1975). Fatty acid composition of the plasma lipids in Greenland Eskimos. Am. J. Clin. Nutri. 28, 958—966.

11. Dyerberg, J., Bang., H.O, Hjorne, N. (1997). Plasma cholesterol concentration in Caucasian Danes and Greenland Eskimos. Dan. Med. Bull. 24,52-55.

12. Dyerberg. J, (1982). Observations on populations I Greenland and Denmark. Nutritional Evaluation of Long- Chain Fatty Acids in Fish oil. Edi. S. Barlow and M. E. Stansby. Academic Press. New York.

13. E. Bates "Eicosanoids, fatty acids and neutrophils: their relevance to the pathophysiology of disease" Prostaglandins, Leukotrienes,Fatty Acids, vol. 53 pp.75–86, 1995.

14. Elizabeth Mannie. 2006. The truth about Omega-3. Prepared foods. June 2006. Pp 1-7.

15. Farlex Partner Medical Dictionary: dementia. (n.d.) Farlex Partner Medical Dictionary. (2012). Retrieved April 26 2018 from https://medical-dictionary.thefreedictionary.com/dementia.

16. Genuis S. J. and Schwalfenberg G. K. Time for an oil checkup: The role of essential fatty acids in maternal and pediatric health. J, Perinatal. 2006: 26(6) 359-65.

17. Gerster H (1998). "Can adults adequately convert alpha-linolenic acid (18:3n-3) to eicosapentaenoic acid (20:5n-3) and docosahexaenoic acid (22:6n-3)?". Int. J. Vitam. Nutr. Res. 68 (3): 159–73. PMID 9637947.

18. Goldblatt MW (May 1935). "Properties of human seminal plasma". J Physiol. 84 (2): 208–18. PMC 1394818/ . PMID 16994667.

19. Gopakumar, K. 1997. Biochemical Composition of Indian Food Fishes. Second Edition, CIFT, Cochin 29.

20. H. Katsuki and S. Okuda "Arachidonic acid as a neurotoxic and neurotrophic substance" ProgNeurobiol, vol. 46(6) pp. 607-636, 1995.

21. Hatem Tallima and Rashika El Ridi. Arachidonic acid: Physiological roles and potential health benefits. A review. Journal of Advanced Research, Available online, 24 November 2017. https://doi.org/10.1016/j.jare.2017.11.004

22. Hibbeln, J.R and Davis, J.M. 2005. Considerations regarding neuropsychiatric nutritional requirements for intake of omega-3 highly unsaturated fatty acids. Prostaglandins, Leukotrienes and Essential fatty acids. 81(2): 179-156.

23. Hibbeln, J.R. (2002) Seafood consumption, the DHA content of mothers' milk and prevalence rates of postpartum depression: a cross national, ecological analysis. J. Affect. Disord. 69, 15–29.Crossref CAS PubMed Web of Science®Google Scholar.

24. J.-Y. You, C. Peng, X. Liu, X.-J. Ji, J. Lu, Q. Tong, *et al.*Enzymatic hydrolysis and extraction of arachidonic acid rich lipids from Mortierella alpine.Bioresour Technol, 102 (10) (2011), pp. 6088-6094.

25. Jensen DV, Andersen KB, Wagner G. Prostaglandins in the menstrual cycle of women. A review. Dan Med Bull. 1987 Jun;34(3):178-82.

26. Keith L.Black, Brenda Culp , Diane Madison , Otelio S.Randall and William E.M. Lands. The protective effectsof dietary fish oil on focal cerebral infarction. Prostaglandins and Medicine.Volume 3, Issue 5, November 1979, Pages 257-268.

27. Khan, R.S. *et al.* L-hydroxytryptophan in treatment of anxiety disorder. Journal of Affective Disorders, 8: 197-200, 1995.

28. Kwak SM, Myung SK, Lee YJ, Seo HG (2012-04-09). "Efficacy of Omega-3 Fatty Acid Supplements (Eicosapentaenoic Acid and Docosahexaenoic Acid) in the Secondary Prevention of Cardiovascular Disease: A Meta-analysis of Randomized, Double-blind, Placebo-Controlled Trials". Archives of Internal Medicine. 172 (9): 686–94.

29. L. Taber, C.H. Chiu, J. Whelan. Assessment of the arachidonic acid content in foods commonly consumed in the American diet. Lipids, 33 (12) (1998), pp. 1151-1157.

30. Lawrence Levine and Iftekhar Alam. Arachidonic acid metabolism by cells in culture: analyses of culture fluids for cyclooxygenase products by radioimmunoassay before and after separation by high pressure liquid chromatography. Prostaglandins and Medicine.Volume 3, Issue 5, November 1979, Pages 295-304.

31. Llorente, A.M., Jensen, C.L., Voigt, R.G., Fraley, J.K., Berretta, M.C. & Heird, W.C. (2003) Effect of maternal docosahexaenoic acid supplementation on postpartum depression and information processing. Am. J. Obstet. Gynecol. 188, 1348–1353.Crossref CAS PubMed Web of Science®Google Scholar

32. Logan AC. (November 2004). Omega-3 fatty acids and major depression: a primer for mental health professional. Lipids in Health and Disease. 3: 25.doi:10.1186/1476-511X-3-25,PMID 15535884.).

33. Lohner, S.; Fekete, K.; Marosvölgyi, T.; Decsi, T. (2013). "Gender differences in the long-chain polyunsaturated fatty acid status: systematic review of 51 publications.". Ann Nutr Metab. 62 (2): 98–112.

34. Lovington. M. B. Omega-3 fatty acids. Am. Farm Physician. 2004.70. 133-40.

35. M. Zhu, P.P. Zhou, L.J. YuExtraction of lipids from Mortierella alpina and enrichment of arachidonic acid from the fungal lipids.Bioresour Technol, 84 (1) (2002), pp. 93-95.

36. MacLean CH, Newberry SJ, Mojica WA, Khanna P, Issa AM, Suttorp MJ, Lim YW, Traina SB, Hilton L, Garland R, Morton SC (2006-01-25). "Effects of omega-3 fatty acids on cancer risk: a systematic review.". JAMA: The Journal of the American Medical Association. 295 (4): 403–15.

37. Malmsten C L Prostaglandins, Thromboxanes, and Leukotrienes in inflammation. Am. J. Med. 1886 Apr. 28;80 (4B) 11-7.

38. Meharban Singh. Essential fatty acids, DHA and human brain, The Indian Journal of Pediatrics, 2005, Volume 72, Number 3, Page 239.

39. Miller PE, Van Elswyk M, Alexander DD (July 2014). "Long-chain omega-3 fatty acids eicosapentaenoic acid and docosahexaenoic acid and blood pressure: a meta-analysis of randomized controlled trials". American Journal of Hypertension. 27 (7): 885–96.

40. Moncada S, andVane JR. Interrelationships between prostacyclin and thromboxane A2. Ciba Found Symp. 1980;78:165-83.

41. Moncade S.,and Vane JR. The role of prostacyclin in vascular tissue. Federation Proceedings(01 Jan 1979, 38(1):66-71).

42. Morris, M.C., Evans, D.A, Tangney, C.C., Bienis, J.L and Wilson, R.S. 2005. Fish Consumption and cognitive decline with age in large community study. Archives of Neurology, 62(12): 1849-1853.

43. Morris, M.C., Evans, D.A., Bienias, J.L., Tangney, C.C., Bennett, D.A., Wilson, R.S., Aggarwal, N. & Schneider, J. (2003) Consumption of fish and n 3 fatty acids and risk of incident Alzheimer disease. Arch. Neurol. 60, 940–946.Crossref PubMed Web of Science®Google Scholar.

44. Odeleye OE and Watson RR. Health implication of the n-3 fatty acids. Am. J Clin. nutr 1991;53; 177-8.).

45. P.C. Calder Dietary arachidonic acid: harmful, harmless or helpful?.Br J Nutr, 98 (3) (2007), pp. 451-453.

46. P.S. Sagar, U.N. Das. Cytotoxic action of cis-unsaturated fatty acids on human cervical carcinoma (HeLa) cells in vitro, Prostaglandins Leukot Essent Fatty Acids, 53 (4) (1995), pp. 287-299.

47. Peet, M and Stokes, C. 2008. Omega-3 fatty acids in treatment of psychiatric disorders. Drugs. 65(8): 1051-1059.

48. Psychoneuroendocrinology 2011 Apr., 36(3). 393-405. Epub. 2011 Ja. Ev. PMID. 21257271(Pub Med- index for Medline).

49. R. Sanchez-Mejia and L. Mucke "Phospholipase A2 and arachidonic acid in Alzheimer's disease" BiochimBiophysActa, vol. 1801(8) pp. 784- 790, 2010.

50. Ramesh, U.N. Das. Effect of cis-unsaturated fatty acids on Meth-A ascitic tumour cells in vitro and in vivo. Cancer Lett, 123 (2) (1998), pp. 207-214.

51. Reeves AM, Austin MP and Parker G., April 2005. Role of omega-3 fatty acid as a treatment for depression in perinatal period. The Australian and New Zealand Journal of Psychiatry 39 (4):274-80.

52. Rizos EC, Ntzani EE, Bika E, Kostapanos MS, Elisaf MS (September 2012). "Association Between Omega-3 Fatty Acid Supplementation and Risk of Major Cardiovascular Disease Events A Systematic Review and Meta-analysis". JAMA. 308 (10): 1024–33.

53. Samuelsson B., An elucidation of the arachidonic acid cascade. Discovery of prostaglandins, thromboxane and leukotrienes. Drugs. 1987; 33 suppl.1:2-9.

54. Sharma JN and Mohammed LA.The role of leukotrienes in the pathophysiology of inflammatory disorders: is there a case for revisiting leukotrienes as therapeutic targets?Inflammopharmacology. 2006 Mar;14(1-2):10-6.

55. Silvers, K.M. & Scott, K.M. (2002) Fish consumption and self reported physical and mental health status. Pub. Health Nutr. 5, 427–431. Crossref PubMed Web of Science®Google Scholar.

56. Singh. M. Essential Fatty acids, DHA and EPA and human brain. Indian J, Pediatr. 2005.,72(3): 239-42.

57. Sirois P and Borgeat P. Leukotrienes. Sem Hop. 1984 Mar 29; 60914): 979-85.

58. Sourairac, A. *et al.* Action of 5-hydroxytryptphan, serotonin precursor, on insomniacs. Annals Medicon Psychologiques, 135: 792-798, 1997.

59. T. Komprda, J. Zelenka, E. Fajmonová, M. Fialová, D. KladrobaArachidonic acid and long-chain n-3 polyunsaturated fatty acid contents in meat of selected poultry and fish species in relation to dietary fat sources. J Agric. Food Chem., 53 (17) (2005), pp. 6804-6812.

60. Tanskanen, A., Hibbeln, J.R., Tuomilehto, J., Uutela, A., Haukkala, A., Haukkala, A., Viinamaki, H., Lehtonen, J. & Vartiainen, E. (2001) Fish consumption and depressive symptoms in the general population in Finland. Psychiatr. Serv. 52, 4. Crossref CAS Web of Science®Google Scholar.

61. Theodore J.Torphy, Bradley J.Undem., Douglas W.P.Hay, Theodore J.Torphy and Bradley J.Undem. Cysteinyl leukotrienes in asthma: old mediators up to new tricks. Trends in Pharmacological Sciences., September 1995, Pages 304-309.

62. U.N. Das. Gamma-linolenic acid, arachidonic acid, and eicosapentaenoic acid as potential anticancer drugs, Nutrition, 6 (6) (1990), pp. 429-434.

63. U.N. Das. Tumoricidal action of cis-unsaturated fatty acids and their relationship to free radicals and lipid peroxidation, Cancer Lett, 56 (3) (1991), pp. 235-243.

64. Van Hiel, L.J. L—5-hydroxytryptophan in depression: the first substitution therapy in psychiatry? The treatment of 99 out patients with therapy- resistant depression. Neuropsychobiology, 6: 230-240, 1990.

65. Von Euler US (1935). "Über die spezifische blutdrucksenkende Substanz des menschlichen Prostata- und Samenblasensekrets"(PDF). Wien Klin Wochenschr. 14 (33): 1182–3. doi:10.1007/BF01778029.

66. Waidwer J. Araragi N, Gutknecht L., Lesch KP.: Tryptophan hydroxylase (TPH2) in disorders of cognitive control and emotion regulation a perspective:

67. X.J. Ji, L.J. Ren, Z.K. Nie, H. Huang, P.K. OuyangFungal arachidonic acid-rich oil: research, development and industrialization.Crit Rev Biotechnol, 34 (3) (2014), pp. 197-214.

68. Yan Cheng, Sandra C. Austin, Bianca Rocca, Beverly H. Koller, Thomas M. Coffman, Tilo Grosse, and John A. Lawson, G.S . Role of Prostacyclin in the Cardiovascular Response to Thromboxane A2. Science 19 Apr 2002: Vol. 296, Issue 5567, pp. 539-541. DOI: 10.1126/science.1068711.

69. Yehuda, S., Rabinovitz, S. & Mostofsky, D.I. (1999) Essential fatty acids are mediators of brainbiochemistry and cognitive functions. J. Neurosci. Res. 56, 565–570.

70. Young, G and Conquer, J. 2005. Omega-3 fatty acids and neuropsychiatric disorders. Reproduction Nutrition Development. 45(11): 1-28.

71. Young. G. and Conquor. J. Omega-3 fatty acids and neuropsychiatric disorders. Reprod.Nut. Dev. 2005 (1): 1-28.

CHAPTER **5**

CHITIN: BIO-MEDICAL APPLICATIONS OF CHITIN AND ITS HYDROLYTIC PRODUCTS

5.1 HISTORY

Chitin is the second abundant polysaccharide occurring in nature. It is also seen in the cell walls of fungi and insects. Crustacean shells contain two important bio-polymers chitin and chitosan.

According Jeuniaux (1961) chitin was first identified by Professor Braconnot, a French Professor of Natural History and Director of Botanical garden, Academy of Nancy, France. He in 1811 isolated it from the fungi called *Fungine*. In 1823, Odier identified it from the cuticle of beetle and given the name "Chitin", a Greek word meaning "coat of mail". In 1843 Payen detected the presence of nitrogen in chitin. In 1876 Ledderhouse hydrolyzed chitin and shoed the presence of glucosamine and acetic acid in the hydrolyzate. The first report of production of chitosan from chitin by hydrolysis with boiling potassium hydroxide was reported in Europe by Rouget in 1889. However, the name chitosan was supposed to be given by Hoppe-Seylor in 1894 according European records. But there is also a popular view in the East that this name was of Japanese origin as the word "san" is a typical Japanese word.

But the final structural elucidation of chitin was done by two schools of investigators, one headed by the celebrated organic chemist and Nobel Laureate, Paul Karrer and co-workers and Zechmeister and his co-workers independently.

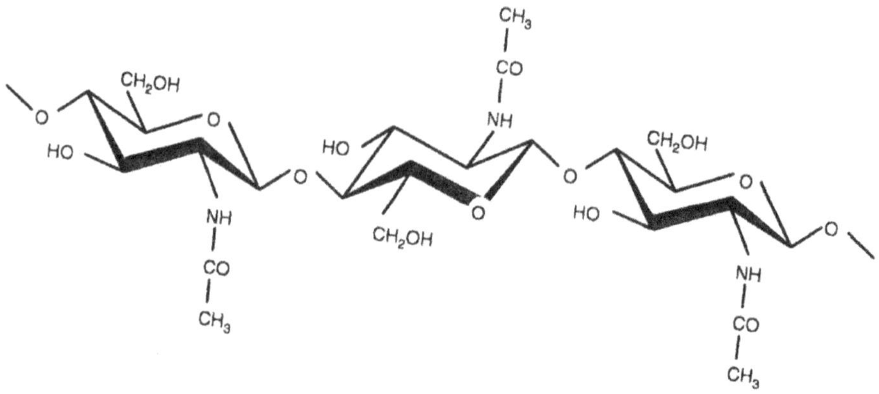

Fig. 17: Structure of chitin

Chitin is the most important constituent of the exoskeletal material of the invertebrates, particularly of shrimps, lobster, squilla and crab. For its abundance in nature it is second only to cellulose. Today they find immense commercial applications. Chitin is a macromolecular linear polymer of anhydro-N-acetyl-D-glucosamine. The monomer units are linked by beta (1-4) glycosidic bonds as in cellulose. Chitin is insoluble in water and most organic solvents. Chitosan is deacetylated chitin.

Fig. 18: Prawn shell waste

Prawn/ shrimp shell is by far the cheapest and most abundant naturally occurring bio-waste suitable for commercial exploitation for the production of chitin and chitosan. Shrimp processing industry, on an average, turns up 40-45% by wet weight as head and hull waste from peeled and de-veined shrimp. If not utilized properly it pollutes the factory premises causing foul smell and becomes an environmental hazard. The proximate composition of prawn waste is given in Table 11 and 12.

Table 11: Proximate composition of shrimp waste

Component	Head and hull waste
Moisture (wet basis) %	76.62
Ash (dry basis,100g dry shell)	31.13
Protein (dry weight %)	39.76
Chitin (dry weight %)	23.08
Fat (dry weight %)	5.05

Table 12: Protein and chitin of prawn shell waste

	Crude Protein %	Chitin %
Head	52 – 55	9 – 11
Hulls	44 – 47	11 – 12

5.2 PRODUCTION OF CHITIN

The crustacean shells contain high quantity of minerals, chiefly calcium, fat and proteins. For the production of chitin these materials have to be removed. The processes are called demineralization and de-proteinisation. De-proteinisation is carried out in steam heated reaction vessels with 3% alkali at 60-70 C for 30 minutes. The de-protienised washed material is de-mineralized with 1.25N hydrochloric acid at room temperature for I hour. All minerals are converted to chlorides and the residue is washed free of traces of acid present. The residue is dried and powdered. The product is chitin. The flow diagram is given in Fig. 19.

5.3 CHITOSAN

Chitin is deacetylated to get chitosan. The chitin is digested with 40% sodium hydroxide at 90-95°C in stainless steel reaction steam heated reaction vessels for 1.5- 2 hours. The process of deacetylation is measured by noting the solubility of chitin in the vessel in 1% acetic acid. One hundred percent solubility is obtained when 90% solubility is obtained. Then the reaction is stopped, alkali drained off, the chitosan is washed free of traces of alkali, dried and powdered. For industrial flocculation applications chitosan of high viscosity is needed whereas for medical application very low viscosity chitosan is needed. For cosmetic applications water-soluble chitosan is used.

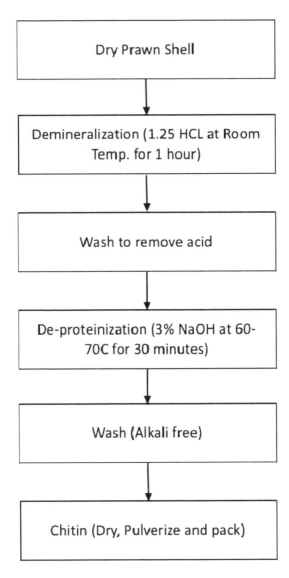

Fig. 19. Manufacturing of Chitin

5.3.1 Production of Chitosan

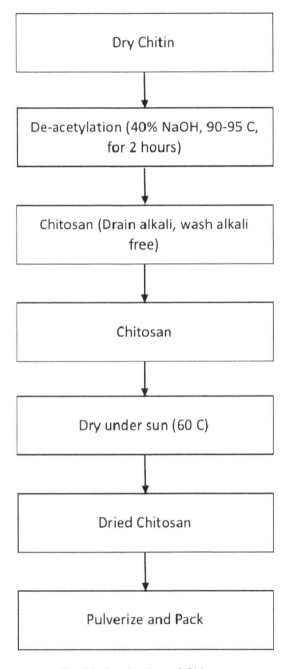

Fig. 20. Production of Chitosan

Chitin and Chitosan plant

Designed and set by, K. Gopakumar at Roxas, Philippines for FAO, 1995

Fig. 21: Chitosan plan

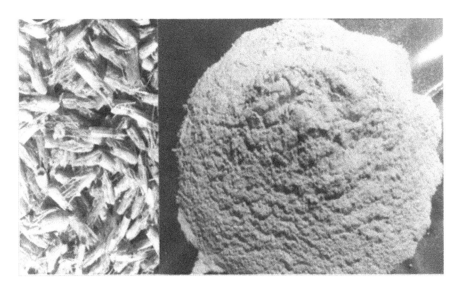

Fig. 22: Prawn shell and finished product, chitosan

Fig. 23: Chitin production and packing
(Courtesy, India seafoods Kochi)

Fig 24: Chitosan ready for export at India Seafoods Kochi

5.4 INDUSTRIAL APPLICATIONS OF CHITOSAN

- **Agriculture:** Regulator of pathogen infection, seed coating, prevents ripening of fruits, anti-fungal, anti- nematode agent, Anti-viral agent, slow release of pesticides, fertilizers, insecticides etc.

- **Research:** As a supporter in chromatography, removes heavy metals by chelation, bio-membranes, enzyme immobilizer, and removal of proteins by absorption, in ultra-filtration and reverse osmosis, flocculation and precipitation

- **Food industry:** As an emulsifier, deacidification of coffee, For clarification of wine and beverages, Substrate for flavor production, For texturization of binder etc. for processed foods, stabilizer and carrier of enzymes, As an emulsifier, deacidification of coffee, for clarification of wine and beverages, Substrate for flavor production, for texturization of foods, As stabilizer, thickener, binder etc. for processed foods, stabilizer and carrier of enzymes

- **Cosmetics:** As moisturizer in skin care cosmetics, adhesive and film forming agent for hair sprays, emulsion thickener& stabilizer for creams, wound healing agent in cosmetics, Increases viscosity of shampoos, creams, gels etc.

- **Biotechnology:** manufacture of drugs and immobilizer and stabilizer of enzymes, cell immobilization, matrix for cell culture, selective recovery of single cell proteins and micro-encapsulation feeds

- **Medical and health care:** Accelerator of healing wounds, ulcers, burns, manufacture of hard and soft contact lens, biodegradable surgical sutures, artificial skin for burns, a material for regeneration of tissues, regenerator of tissues in ortho- dental surgeries, slow release of drugs, anti- tumour agent, bacteriostatic agent, membranes for haemodialysis, prevents gastric acidity, immunoadjuvant hyper-cholesterolmic agent and also as a haemostatic agent.

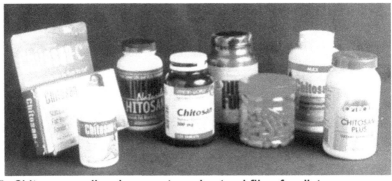

Fig. 25: Chitosan as slimming agents and natural fiber for dietary use-commercial products

- One of the important uses of chitosan is now as a slimming agent. Chitosan is a natural fiber. Taken after diet it binds fat and make it non-available for absorption in the intestines and then excreted. There are a number of commercial chitosan based slimming agents available in the market.

- **Textile and paper industry:** To lower the density of silver, as a fixative, as protective coating, Water repellent, improves wet and dry strength, as a sizing agent in paper industry, Improves dye binding in textile industry, for retention of acidity

- **Industrial flocculent:** Waste water treatment, forms insoluble complexes with several dyes, Complexes tannins, removes toxic metal like Hg, Pb, Cd, Zn etc used in metallurgy for separation of ores. The prawn shell an environmental hazard to seafood industries has become a useful by-product of great value.

Biotechnological applications of chitins

- Chitin and chitosan find extensive applications in bio-technological and biomedical uses. Some of them are listed below:

- Controlled release of drugs, flavours and antioxidants.

- Control of enzymatic browning in fruits and vegetables.

- Reduction of oxygen partial pressure.

- Immobilization of enzymes

- Manufacture of membranes for reverse osmosis

Chitosans are degraded by enzymes in nature. Certain bacterial cell walls contain enzymes that hydrolyze both chitin and chitosan (Berger and Weiser, 1957). Hence, they find application in agriculture to improve soil condition and infestation by nematodes.

5.5 USE IN HUMAN HEALTH

Today chitosan is extensively used as a slimming agent to prevent obesity in man and animals. When used along with diet as a supplement chitosan chelates fat in food excreted by man and there by prevent absorption by human body. There are now several formulations sold in the market as a slimming agent. Oral administration of chitosan reduces serum cholesterol levels and brings weight loss (Macchi, 1996, Colombo and Sciatto, 1996, Sugara et al. 1978. Pitter and Abbot, 1999., Bokora and Kabayashi 2003. 2003). Chitosan film is also applied for medical uses (Hosokawa et al. 1991.)

In human body chitin is not present as such. But its hydrolytic products, N-acetyl glucosamine and N- glucosamine are present (Muzzarelli, 1997). But there are several enzymes, hydrolases, present in human body, that can breakdown chitin. Among them the most important one is the lysozyme (Despandey, 1986). These enzymes cause biodegradation in vivo in humans. There are now evidences to suggest that human chitinases are present. Such activity was reported to be present in human serum and leucocytes (Escott and Dams, 1995). One of the important study reported on biodegradation of chitin and chitosan by enzymes was the work of Tomihata and Ikara (1997). They found that the degradation of chitosan lysozyme is complete if deacetylation is below 70% whereas chitosan having deacetylation above 70% hydrolysis is poor and pure chitosan is not hydrolyzed at all. This means that for medical applications where biodegradability is desired chitosan having low deacetylation is the ideal one. This finding was in agreement with the work reported by Central Institute of Fisheries Technology, Kochi (Gopakumar, 1997.) that chitosan of low viscosity, less than 10 cp are ideal for medical application. Very high viscosity chitosan has the highest degree of deacetylation, above 85%, and are ideally suited for use as flocculating agent for industrial uses like purification of industrial effluents, removal of heavy metals by chelation, purification of wine etc.

Chitin is a biopolymer having great compatibility with animal tissues. Chitin and chitosan can be used both as a bio-stable or bio-degradable material for medical use, A biomaterial is defined as:

"non-viable material used in a medical device, intended to interact with biological systems." (Williams, 1987). There are now several medical products available in the market made of / containing chitosan like wound dressings and as blood contact tubing. They fall under bio-stable category. But bone and tissue substitutes used come under bio-degradable items.

5.6 CHITOSAN DERIVATIVES

Chitosan based several products are now seen the markets. The presence of highly reactive primary amino groups in chitosan is the backbone responsible for the reactivity that enable chemists to synthesize a perplexing array of derivatives like water soluble chitosan, amino acid-modified chitosan and other synthetic derivatives. Modification of amino groups improves properties like water solubility and mucoadhesiveness needed for biomedical applications. Water soluble chitosan finds extensive applications in cosmetic industries for the formulation of creams and lotions.

The introduction of amino acid moieties to the backbone of chitosan has been shown to improve its physic-chemical properties and gives rise to some interesting synergistic characteristics for use in drug delivery and tissue engineering, antimicrobial, anticoagulant and anti-cholesterol activities (Luca Casettari *et al.* (2012).

Conclusions: Biodegradability of chitins and chitosan are still controversial. There are many limitations. The end products of its degradation products should be harmless. This is a primary requirement as far as safety of the biomolecule is concerned. Still more investigations are needed to clear the bio-safety of chitin and its derivatives

2.7 GLUCOSE AMINE

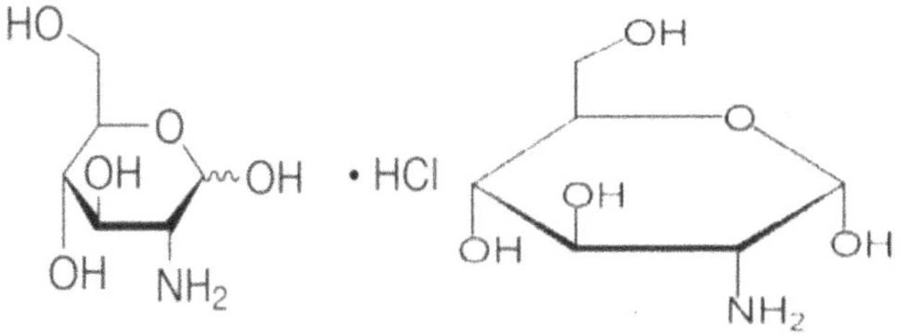

Fig. 26. D-(+)-Glucosamine hydrochloride ($C_6H_{13}NO_5$. HCl)

D-glucosamine ($C_6H_{13}NO_5$) or2-amino-2-deoxy-D-glucose is an amino sugar (hexosamine) with a molecular weight of 179.17. It is naturally present in human body and crustacean shells. Glucosamine is a precursor of biochemical synthesis of the GAGs (glycosaminoglycans) that is found in cartilage. Premature loss of cartilage is part of the clinical syndrome recognized as OA (osteoarthritis). (Clegg, D. O., and Jackson, Ch. G. 2005).

Glucosamine in the form of glucosamine sulphate,glucosamine hydrochloride, or N-acetyl-glucosamineis are now extensively used as a dietary supplement in the treatment for osteoarthritis, knee pain, and back pain. Clinical studies indicated that glucosamine is safe under current conditions of use and does not affect glucose metabolism (Houpt, 1990; Luo, 2005)

Acid hydrolysis of chitin gives a valuable chemical now widely used in pharmaceutical and chemical industries called glucose amine. Glucoseamine is produced from chitin by hydrolysis with concentrated hydrochloric acid under controlled conditions in glass-lined reaction vessels. This chemical is now in great demand and also very expensive. The demand for glucoseamine hydrochloride is very high from pharmaceutical industries as it can be used as slow releaser of drugs.

Both glucosamine hydrochloride and glucosamine sulphate are used for medical applications especially for the treatment of osteoarthritis and other inflammatory disorders of joints in human body. Of the two, the sulphated form is better for therapeutic applications as proved by clinical trials.

5.8 METHOD OF PRODUCTION OF GLUCOSE AMINE (CIFT PROCESS)

Hydrolysis of chitin with 12 M HCl is done in an all glass reaction vessel. Steel or other metal reactors unsuitable for acid hydrolysis. The major steps involved in the production of Glucoseamine-HCl from chitin are:

1. Acid hydrolysis of polysaccharide;

2. Filtration of the solution,

3. Recrystallization of the product, and

4. Filtration, washing and drying of final product at 50°C

The temperature of the hydrolysis is 80-90^{0}C and time 90 to 120 minutes. For recrystallization methanol is preferred over ethanol in view of the price factor. The details production is given in the flow diagram Fig. 27.

5.9 REFERENCES

1. Amano KI, Ito E. 1978. The action of lysozyme on partially hyydracetylated chitin. Eur. J. Biochem. 85(1) 97-104.

2. Berger LR, Weiser RS. 1957. The glucosaminidase activity of egg white lysozyme. Biochem. Biophys. Acta26(3): 517-521.

3. Bokura H, Kobayashi S. 2003. Chitosan decreases total cholesterol in Women, a randomized, double-blind, placebo- controlled trial. Eur. J. Clin. Nutr 57: 721-725.

4. C. Jeuniaux. 1961. An addition to the list of hydrolysase in the digestive tract of vertebrates. Nature 192: 135-136.

5. Casettari L, Vllasaliu D, Lam JK, Soliman M, Illum L. 2012. Biomedical applications of amino acid – modified chitosans: A review. Biomedicals 33(30): 7565-7583.

Production of glucosamine from shrimp shell

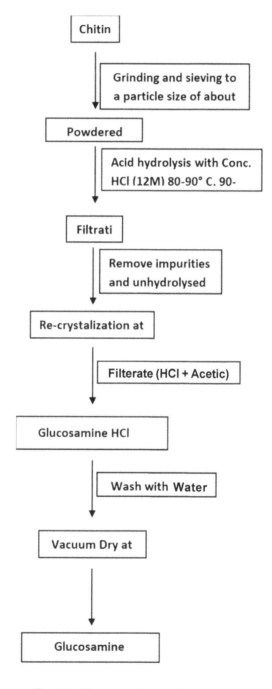

Fig. 27: Glucosamine production

6. Clegg, D. O., and Jackson, Ch. G. 2005. "Glusosamine." In Encyclopedia of Dietary Supplements, edited by Coates, P. M., Blackman, M. R., Cragg, G., Levine, M., Moss, J., and White, J. D. New York: Marcel Dekker

7. Colombo P, Sciatto AM.1996. Nutritional aspects of chitosan employment in hypocaloric diet. Acta. Toxicol. Therap. 17: 287-302.

8. Despande MV. 1986. Enzymatic degradation of chitin and its biological application. J. Science and Industrial Research 45:273-281.

9. Escott GM, Adams DJ. 1995. Chitinase activity in human serum and leukocytes. Infection and immunity. 63(12): 4770-4773.

10. Gopakumar. K Tropical Fishery Products. Oxford and IBH Publishing Company Pvt Ltd New Delhi. 1997.

11. Hosokawa J,Nishiyama M, Yoshihara K, Kubo T, Terabe A. 1991. Reaction between chitosan and cellulose on biodegradable film formation: Ind. Eng. Chem. Res 30(4): 788-792.

12. Houpt, J. B., McMillan, R., Wein, C., and Paget-Dellio, S. D. 1999. "Effect of Glucosamine Hydrochloride in the Treatment of Pain of Osteoarthritis of the Knee." J. Rheumatol. 26: 2423-30.

13. Illum L. 1998. Chitosan and its use as a pharmaceutical excipient. Pharm Res 15(9): 1326.

14. Jeuniaux C. 1996. A brief survey of the early contribution of European scientists to chitin knowledge. In: Domard A, Jeuniaux C, Muzzarelli RAA,

15. Roberts G, editors. Advances in Chitin Sciences, Jacques André Publishers, Lyon, France 1996; 1–9.

16. Luo, J., Hu, Y. S., Wu, Y., and Fan, W. K. 2005. "Effect of Glucosamine Hydrochloride in Ameliorating Knee Osteoarthritis." Chin. J. Clini. Rehabil. 9: 70-2.

17. Macchi G. 1996. A new approach to treatment of obesity: chitosans's effect on body weight on body weight reductions plasma cholesterol levels. Acta. Toxicol. Therap 17: 303-320.

18. Muzazarelli RAA. 1997. Human enzymatic administration of chitin derivatives. Cellular and Molecular Life Sciences. 53: 131-140.

19. Muzzarelli RA, Mattioli-Belmonte M, Pugnaloni A, Biagini G. 1999. Biochemistry, histology and clinical uses of chitin and chitosan in wound healing. Exs. 87:251-64.

20. Pittler MH, Abbot NC, Harkness EF, Ernst E.1999. Randomized double-blind trial of chitosan for body weight reduction. Eur. J. Clin. Nutr53(5):379-81.

21. Richards AG. The composition of chitin and cuticle in Nature. In the integuments of arthropods: The chemical components and their properties, the anatomy and development, and the permeability, University of Minnesota Press. Minneapolis MN. 1951.54.

22. Tomihata K, Ikada Y. 1997. In vitro and In vivo degradation of films of chitin and its deacedylated derivatives. Biomaterials 18(7):567-575.

23. Williams DF. Definition in Biomaterials. 1987. Proceedings of a consensus conference of European Society of Biomaterials. Chester, England. March. 3-5, 1986 Vol. 4. Elsevier, New York.

CHAPTER **6**

SEA CUCUMBERS

S ea cucumbers are invertebrate animals found both in tropical and subtropical oceans. They are habitats of inter-tidal zones and deeper waters. Sea cucumbers are characterized by cylindrical or tubular structure resembling very much to cucumbers and hence, got the name sea cucumber. There are more than 1100 species listed under this group. Sea cucumbers belong to the class Holothuroidea in the phylum *Echinodermata (*meaning spiny-skinned) and divided into three subclasses namely *Dendrochirotacea, Aspidochirotacea*, and *Apodacea.* Under these subclasses there are six orders named as *Aspidochirotida, Apodida, Dactylochirotida, Dendrochirotida, Elasipodida and Molpadiida* (Connad 2004, Ridzwan 2007). The name holothuroid was given by the Greek philosopher, Aristotle ("holos: whole" and "thurios: rushing). The scientific name of sea cucumber, *"Cucumismarimus"*, was coined by an invertebrate taxonomist Pliny (Fell 1972). Their Latin name is *Stichopus chloronotus.* Sea cucumbers are widely used as food delicacy in all South east Asian countries. It is also called trepang or balata.According to publications in Chinese traditional medicines (Tang Weici, 1987.; Zhang Enchin 1988) sea cucumbers can improve human immune status enforcing resistance to many diseases and even have anticancer properties. Chinese name for sea cucumber is *haishen* meaning "ginseng of the sea" because ginseng, a plant belonging to the family *Araliaceae*, has similar medicinal properties. The sea cucumber is valued-along with several other delicacies, such as shark's fin, ginseng, cordyceps, and tremella-as a disease preventive and longevity tonic. It was listed as a medicinal agent in the **Bencao Congxin (New Compilation of Materia Medica**) by Wu Yiluo in 1757. It is often known in medical literature as *fangcishen (fang* = four-sided, *ci* = thorny; referring to the spiky protrusions that emanate from four sides) or, in abbreviated form, *fangshen* (Subhuti Dharmananda).

Sand fish (*Holothuria scabra*)

Holothuria scabra and H. scabra var. versicolor (now H. lessoni) have recently been properly identified as distinctive species.Holothuria scabra is occurring in the East coast of India, in the Gulf of mannar and other sea coast. It is also occurring in Australia; Bangladesh; British Indian Ocean Territory; Brunei Darussalam; Cambodia; China; Comoros; Cook Islands; Djibouti; Egypt; Eritrea; Fiji; India (Andaman Is.); Indonesia; Israel; Japan; Jordan; Kenya; Kiribati; Madagascar; Malaysia; Maldives; Mauritius; Mayotte; Micronesia, Federated States of ; Mozambique; Myanmar; Nauru; New Caledonia; Oman; Palau; Papua New Guinea; Philippines; Réunion; Samoa; Saudi Arabia; Seychelles; Singapore; Solomon Islands; Somalia; South Africa; Sri Lanka; Sudan; Taiwan, Province of China; Tanzania, United Republic of; Thailand; Tonga; Tuvalu; Vanuatu; Viet Nam; Wallis and Futuna; Yemen (Ref.FAO Marine Fishing Areas).

The well-known Indian species widely distributed in the East coast of India and Andaman and Nicobar Islands is Holothuria scrab a popularly known as sandfish. In India, Holothuria *scabra*, H. spinifera and Bohadschia marmorata have been collected over the last 1000 years. Fishermen started hrvesting other species by 1990's because of high export value and population declines. *Actinpyga echinites* and *A. miliaris* species are now overexploited in some areas particularly in the Gulf of Manner and Pal Bay areas. CPUE and size of specimens have dramatically declined. Harvesting and export of sea cumber is now banned by an act by Government of India since 2008 in the Andaman and Nicobar Islands and also East coast of India. Still it is caught and exported to some Asian countries by clandestine traders particularly from Andaman and Nicobar Islands coast.

Beche-de-mer

The processed form of sea cucumber is well known as Beche-De- Mer (literally "sea-spade"). It has different name in different countries. In French, it is called *trepang* (or *trîpang*), in Indonesian, *namako*, in Japanese, and *balatan* in Tagalog. In Malay, it is known as the *gamat* (Alessandro Lovatelli, and C. Conand, 2004). Beche-de-mer is mainly used to make soup in South/eastern countries of Asia. It is now an expensive delicasy.

Some of the commercially important species in Asia- Pacific oceans are listed in Table 13.

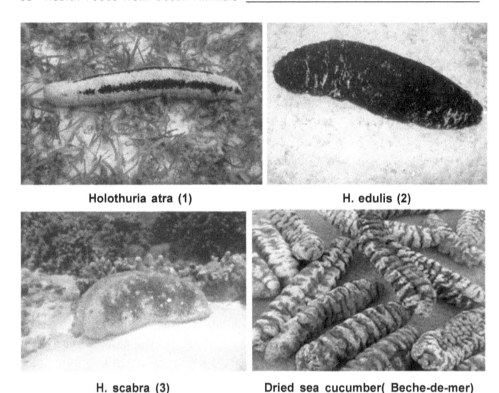

Holothuria atra (1)   H. edulis (2)

H. scabra (3)   Dried sea cucumber(Beche-de-mer)

Fig 28: Picture - Sea cucumber (Holothuria atra, edulis, scabra and dried sea cucumber)
Photos 1,2,3, courtesy Dr. K K. Idrees Babu

6.1 PROXIMATE COMPOSITION OF SEA CUCUMBERS

It is very difficult to give a standard proximate composition of sea cucumber as there is wide variation between species to species. Even in a single species biochemical composition can have wide fluctuation depending on feeding status, seawater temperature, age, seasonal variation, breeding status etc. At best biochemical composition can only be expressed in arrange of values for major components. Sea cucumbers contain generally high amount of moisture, and lower amount of proteins and lipids compared to fish and shellfish. They contain relatively high amount of fiber compared to fish(Ali Raza *et al.*, 2012.). The edible wall is a rich source of collagen and more than 21 types of collagen has been identified so far (Gelse *et al.*, 2003).Sea cucumber has 10-16% mucopolysaccharides (mainly chondroitin sulfate) building blocks of the cartilage. Chondroitin sulphate in combination with glucosamine is now extensively used to treat osteoarthritis.

Table 13: Sea cucumber species of commercial importance seen in Tropical seas.

Name of fish	Scientific name	Length (cm)	Width (cm)	Body wall thickness, (mm)	Live weight, (kg)
Black teat fish	*Holothuria nobilis*	30 – 40	10 – 15	10 - 12	2 – 3
White teat fish	*Holothuria fuscogilva*	30 -40	10 – 15	10 – 12	2 -3
Prickly redfish	*Thelenotamananas*	40 – 70	10 – 15	15 - 20	1 – 6
Deep-water redfish	*Actinopygaechinites*	15 – 30	8 – 10	7	0.5 – 1
Stonefish	*Acinopygalecanora*	Up to 40	9 –	—	—
Surf redfish	*Acinopygamauritina*	20 – 30	8 – 10	6	0.5 - 1
Blackfish	*Acinopygamiliaris*	20 – 30	8–12	8	0.5 – 2
Brown sandfish	*Bohadschiamarmorata/vitiensis*	15 – 35	8 – 10	5 – 10	0.5 – 2
Lolly fish	*Holothuria atra*	10 – 50	3 – 8	2 – 6	0.2 – 1.5
Pink fish	*Holothuriaedulis*	20 – 30	3 – 6	2 – 5	0.2- 0.4
Elephant's Trunk fish	*Holothuriafuscopunctate*	40 – 60	10 – 15	8 – 12	2 – 4
Green fish	*Stichopuschloronotus*	10 – 30	3 – 6	2 – 6	0.2 – 0.4
Curry fish	*Stichopusvariegatus*	20 – 25	6 – 12	6 – 10	0.8 – 2
Amber fish	*Thelenataanax*	Up to 80	Upto 15	—	—
Name of fish	*Holothurianobilis*	30 – 40	10 – 15	10 - 12	2 – 3
Black teat fish	*Holothuriafuscogilva*	30 -40	10 – 15	10 – 12	2 -3
White teat fish	*Thelenotaananas*	40 – 70	10 - 15	15 - 20	1 – 6

Table 14: Proximate composition of Sea cucumbers
Proximate composition. g/100 g wet whole body)

Composition	%
Moisture	82 – 92%
Protein	2.5 – 13.8%
Lipids	0.1 – 0.4 %
Carbohydrates	0 – 2.2%
Ash	1.5 – 4.3%
Fiber	0.5 – 2.5%

Sea cucumbers are seen in a variety of colours having length varying from two centimeters to two meters. While some of them are filer feeders, most of the species feed on detritus, planktons and algae. Their body contains feet those are tubular in shape. During idling or resting these tubes are held close to the body. For movements these feet are extended or contracted by adjusting body fluid pressure In Indian coast the most important species commercially used for processing is Holothuria scabra. It is seen to reach a length of 40 cm. The life span is about 10 years. In 18 months this species reaches sexual maturity and breeds twice in a year, March to May and November-December. The sea cucumbers are bottom dwellers. They are usually obtained as by catches in shrimp trawls and also harvested by collection from beaches and shallow waters and by divers from sea bottoms. It is marketed in a dried form called Beche-de-mer.

6.2 MEDICAL USES

Sea cucumbers have been used for therapeutic applications for centuries by people of Asia, particularly the Chinese and Malaysians. A number of ointments and tonics are prepared from sea cucumbers. Initially a decoction is prepared by boiling them in a vat (Fig. 29). Various herbs are also added, and a large number of formulations are available in the market. There is a flourishing industry in Malaysia. None of these remedies are supported by detailed clinical studies. It is a case for reverse pharmacological research and there is scope for more research in this area. In Malaysia there is a flourishing industry producing a variety drugs for treating various human diseases (Gopakumar, 1997).

Sea cucumber whole and dried forms are used as a food or tonic to attain longevity and also as an aphrodisiac. The Chinese drug industry chiefly uses it for treatment of general debility, impotency, anti-aging, to relieve constipation due to intestinal dryness, frequency in urination and arthritis. Recent studies have shown that it is an excellent material for treating Rheumatoid arthritis, Osteo arthritis, Ankylosing spondylitis and several connective tissue disorders. Many people use it also to prevent or control cancer. But none of these properties are supported by concrete clinical trials. These formulations are used only as

Fig 29: Sea cucumber extraction in a vat.

Fig. 30: Sea cucumber extract containing pharmaceutical preparations in a Malaysian factory

unprescribed drugs. Most probably the therapeutic effect on arthritis can be attributed to the chondroitin sulphate and other long chain polysaccharides seen in sea cucumbers. None of these remedies are supported by detailed clinical studies. It is a case for reverse pharmacological research and there is scope for more research in these areas. Most of the preparations are not prescribed by medical professionals but purchased by people based on recommendation of users. But there is a lucrative market for these preparations indicating that results are very positive. Sea cucumber dried, often mixed with herbs, are sold in encapsulated form.

Sea cucumber lipids particularly lipids from gonads contain very high amounts of Poly Unsaturated fatty acids (PUFA), as high as 40-45% of total fatty acids. Lipids in sea cucumber were dominated by eicosapentaenoic acid (EPA, C20:5n-3), ranging from 43.2 to 56.7% of the total fatty acids. Docosahexaenoic acid (DHA, C22:6n-3) was present at a much lower concentration of 2.0-5.81% in the total fatty acid pool (Zhong, Khan and Shahidi 2007). The oil fraction contained a mixture of branched chain fatty acids, 12-MTA (methyltetradecanoic acid) and its isomer compound, 13-MTA (Yang Peiying *et al.*, 2003) Fig. 31. Both are found potent inhibitors of the 5-LOX (lypoxygenase) enzyme system. 5-LOX inhibitors are now used to treat asthma, ulcerative colitis, and arthritis.

Fig 31: 12- methyltetradecanoic acid
(12 Methyl Myristic acid)

Triterpene glycosides

These are compounds composed of two structural components, triterpenes and glycosides.

Triterpenes: They are composed of three molecules of monoterpene. Monoterpenes consists of two isoprene units (2- methyl-1,3 Butadiene, $CH_2=C(CH3)-CH=CH_2$). They can be either cyclic (ring form) or acyclic(liner). One simple monoterpene is Myrcene, having the formula $C_{10} H_{16}$.

Triterpenes have a common molecular formula $C_{30} H_{48}$ and composed of three molecules of monoterpenes. They also exist as a straight linear molecule (e.g. Squalene) or in cyclic form and made up of 6 isoprene units. In cyclic form they exist as a penta cyclic ring (e.g. Hopane).

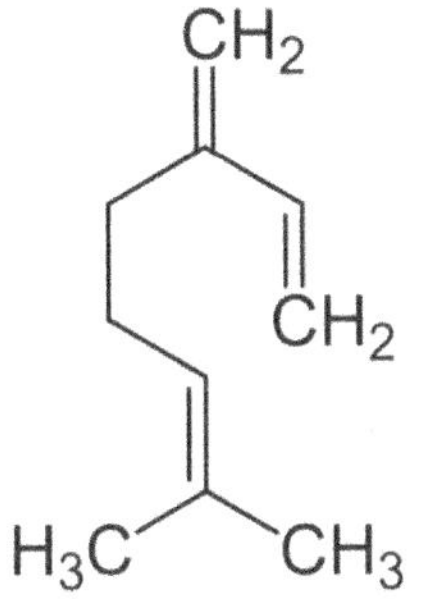

Fig. 32: Myrcene (monoterpene)

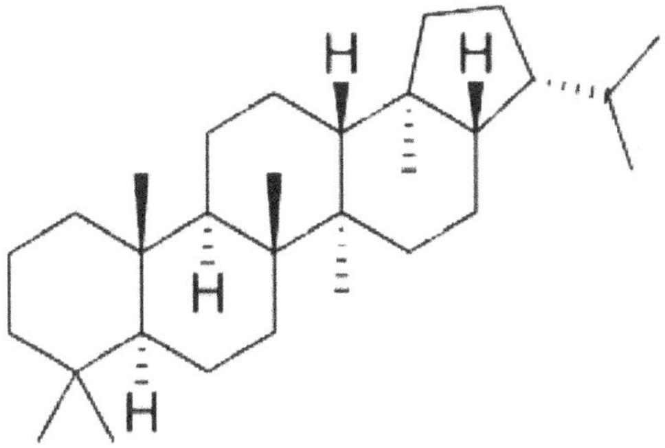

Fig. 33: Structure of Hopane

When functional groups like alcohol (OH), carboxylic (COOH), ketonic (-CO-) etc. are added to them they are often referred to as triterpenoids (e.g. cholesterol).

When sugar molecules are attached to the functional group of another organic molecule eg. triterpenes (triterpenoids) they are called glycosides. There are different types of glycosides like alcoholic glycosides, flavanoid, steroid, phenolic etc. Triterpene glycosides are also referred to as saponins.

The triterpenes are subdivided into 20 subgroups depending on their chemical structure. The number of sugar molecules attached are also different. This explains the reason for the existence of a large number of triterpene glycosides in nature, both in plant and animal kingdom. The common sugar molecules seen are glucose, fructose, xylose and glucuronic acid. Sometimes non-sugar molecules are also seen. Then sugar group is called glycone and non-sugar group as aglycone. The sugar group can be a sugar like glucose(monosaccharide) or several groups (oligosaccharide).

Biochemistry of glycosides

Glycosides can play an important role in nutrition and health in man and animals. Normally they exist in an inactive form. They are activated by enzymes present in animals breaking of the sugar part and releasing the other part, chemical entity. Such chemicals are now found to have many therapeutic properties to fight diseases like cancer, arthritis, rheumatism, as adjuvant in vaccine etc. Over the years a large number of research papers have been published on the isolation of bioactive components from sea cucumber and some of them based on species differences and geographical regions are given inTable 15.

Medical properties attributed to sea cucumber: It appears today and refer to available research publications related to medical applications of sea cucumber it is a universal remedy for all known diseases including cancer and for which no single drug is yet developed. Most of these attributes are based on traditional medical practices and none of them have an approved clinical evidence. Some of the diseases for which sea cucumber meat is widely consumed are given below:

Aches & Pains, Anti-aging, Aphrodisiac, Arthritis, Asthma, Blood pressure, Cancer, Constipation, Cuts & Wounds, Diabetes, Digestive system , Gum disease, Heart disease, Hypertension, Immune system, Impotency, Inflammation, Joint pains, Kidney disorders, Psoriasis & Skin problems, Reproductive problems, Stomach ulcers, Viruses.

Anti-Cancer properties

Some sea cucumber triterpene glycosides exhibit pronounced anticancer effects by direct interaction with tumor cells in the sub-cytotoxic range of concentration. Some sea cucumber triterpene glycosides exhibit pronounced anticancer effects. (Dmitry L. Aminin *et al* ,2015; Tian *et al.*, 2005, 2007).

6.3 TRITERPENE GLYCOSIDES

Triterpenoid glycosides are naturally occurring compounds in sea cucumbers. They are responsible for the toxicity of these echinoderms. More than 100 triterpenoid glycosides have been isolated so far.

Most of the holothurians are toxic. Theyare grouped into four main structural categories considering their aglycone structure: 3β-hydroxyholost-9(ll)-eneaglycone skeleton, 3β-hydroxyholost-7-ene skeleton, other holostane type aglycones and nonholostanem aglycone (Kim and Himaya, 2013). Thetriterpenoid glycosides occurring in most sea cucunbers are found to be possessing potential biological activities. Among the biological activities, anticancer activity and antiviral activity are the widely studied areas. The first anticancer properties of the sea cucumber glycoside, holothurin, representing the glycoside fraction of Bahamian sea cucumber *Actinopygaagassizi*) were described in by Nigrelli (1952).

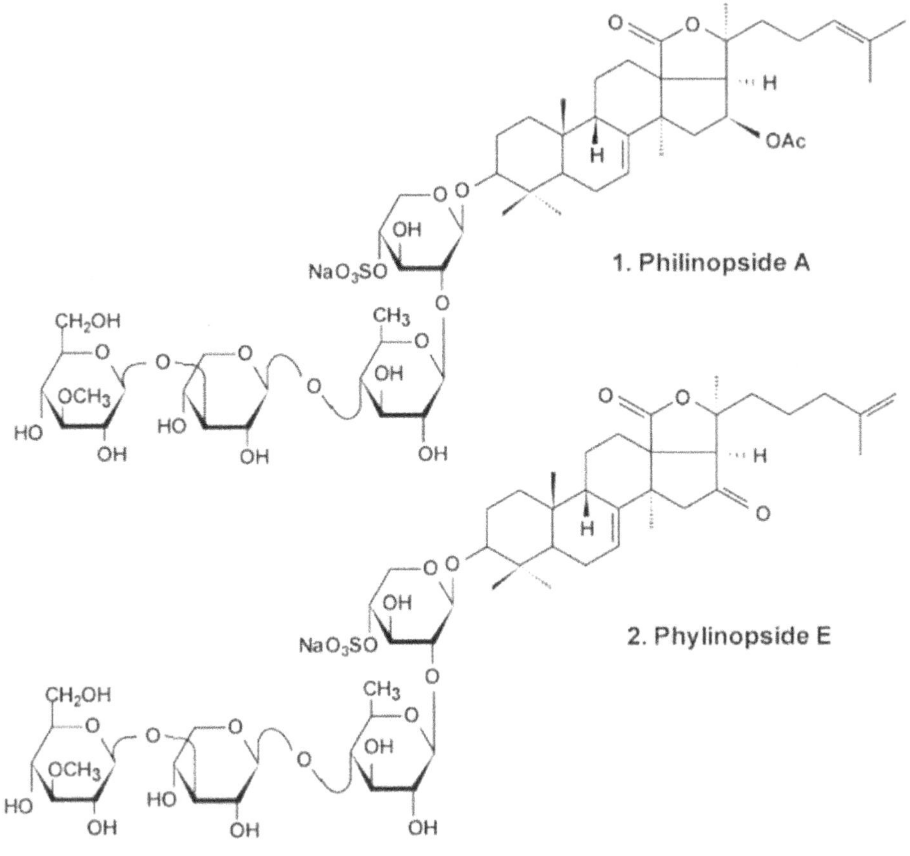

Fig. 34: Triterpene glycosides
Ref. Dmitry L. Aminin *et al.* (2015)

Triterpene glycosides are secondary metabolites of sea cucumbers (*Holothurioidea, Echinodermata*). and to inhibit metastasis and also have hemolytic, cytotoxic, antifungal, and other biological activities. "*Administration of some glycosides leads to a reduction of cancer cell adhesion, suppression of cell migration and tube formation in those cells, suppression of angiogenesis, inhibition of cell proliferation, colony formation and tumor invasion*" (Dmitry L. Aminin *et.al.*, 2015). Yang *et al.* (2003) isolated branched chain fatty acid 12- methyl tetradecanoic acid, from sea cucumber extract and showed its anticancer effect on PC3 prostate cancer cells.

A major challenge for oncologists and pharmacologists is to develop less toxic drugs that will improve the survival of lung cancer patients. Frondoside A is a triterpenoid glycoside isolated from the sea cucumber, Cucumariafrondosa was shown to be a highly safe compound. Investigations of Attoub *et al.* (2012) on the impact of Frondoside A on survival, migration and invasion in vitro, and on tumor growth, metastasis and angiogenesis in vivo alone and in combination with cisplatin and Frondoside A caused concentration-dependent reduction in viability of LNM35, A549, NCI-H460-Luc2, MDA-MB-435, MCF-7, and HepG2 over 24 hours through a caspase 3/7-dependent cell death pathway. The IC50 concentrations (producing half-maximal inhibition) at 24 h were between 1.7 and 2.5 µM of Frondoside A. The authors also observed that Frondoside A induced a time- and concentration-dependent inhibition of cell migration, invasion and angiogenesis in vitro. Frondoside A (0.01 and 1 mg/kg/day i.p. for 25 days) significantly decreased the growth, the angiogenesis and lymph node metastasis of LNM35 tumor xenografts in athymic mice, without obvious toxic side-effects. Frondoside A (0.1-0.5 µM) also significantly prevented basal and bFGFinduced angiogenesis in the CAM angiogenesis assay. Moreover, Frondoside A enhanced the inhibition of lung tumor growth induced by the chemotherapeutic agent cisplatin.

These findings indicate that sea cucumber is a promising therapeutic agent for treating lung cancer.

6.3.1 Antioxidant properties

Althunibat *et al.* (2013) showed that aqueous and organic extracts of *Holothuria edulis* lesson and *Stichopushorrens* Selenka. *Stichopushorrens* contain promising levels of antioxidant and cytotoxic natural products that might be used for cancer prevention and treatment. Rehydrated sea cucumber samples especially with internal organs were found to exhibit higher free radical scavenging property against 2,2'-azobis(2-aminopropane) dihydrochloride (AAPH) and 1,1-diphenyl-2-picrylhydrazyl (DPPH) radicals than their fresh counterparts (Zhong *et al.,* 2007).

6.3.2 Sea cucumber as aphrodisiac

Sea cucumbers has a distinctly phallic appearance, and this may account for its reputation as an aphrodisiac. They contain niacin, magnesium, vitamin C and zinc and that help to reduce muscle tension, increase blood flow, build sex hormones and maintain a healthy sperm count and improves prostate health by scavenging free radicals. Dried sea cucumber powder capsule form is sold in many parts of the world as a substitute to Viagra. But no clinical study by competent research organizations supports this claim. But still it enjoys a lucrative market globally.

Table 15: Sea cucumber species from several parts of the world and compounds isolated

No.	Name of species	Compounds isolated	Reference
1.	*Athyonidium chilensis*	Saponins	Ignacia *et al.* 2013
2.	Okinawan sea cucumber	Triterpeneoligoglycosides	Kobayashi *et al.* 1991
3.	*Dendrochirotida*	Triterpene glycosides	Kalinin *et al.* 1996
4.	*Echinodermatan*	Triterpene glycosides	Chudil *et al.* 2003
5.	*Medicago Arabica*	Triterpenesaponins	Bialy *et al.* 2004
6.	*Holothuriapolii*	Methanolic extracts	Ismail *et al.* 2008
7.	*Cucumaria frondosa*	Lipids: Red oil & Gold oil	Collin& Peter Davil. US Patent. WO 79/87314/1999
8.	*Holothuria Leospilota & S. Japonicus*(gonads)	Esters of Echineone, Xanthins and Beta- carotene	Matsuno *et al.* 1995
9.	*Polusfabrichii* (body walls)	Xanthins	Tsushima y*et al.*, 1996 and Findlay *et al.* 1983.
10.	Atlantic sea cucumber (*Cucumaria frondosa*).	Phenolic compounds	Mamelona *et al.* 2007
11.	*Holothuria forskali*	Nonsulphatedtriterpenoid glycosides	Rodriguez *et al.* 1991
12.	*Cucumaria frondosa*	Antioxidants	Ying Zhong *et al.* 2007
13.	Antractic sea cucumber (*Staurocucumis liouviellei*)	Triterpene glycosides	Alexander *et al.* 2015
14.	*Stichopus japonicas*	Neuritrogenic active galglioside	Masafumi Kaneko *et al.* 1999.
15.	*Astralostichopusmollis*	Uncommon monosulphated triterpene glycosides mollsosides	Moraes *et al.* 2005

[Table Contd.

Contd. Table]

No.	Name of species	Compounds isolated	Reference
16.	*Cucumariafrondosa*	AglyconeFrondogenin-A	Findlay J. A. and Anand D., 1984
17.	*Holothurianobilis*	Triterpene Glycosides, Nobilisides	Jun Wu et al., 2006
18.	*Hemoidemaspectabilis (from Patagonia)*	Triterpene glycosides	Chudil et al. 2002
19.	*Mensamariaintercedens, Lampert.*	Triterpene glycosides A-C, D	Zou etal.,2003, 2005.
20.	*Acoudinamolpadioides*	Fucoidan	HuiXu et al. 2014
21.	*Holothurioidae*	Triterpene glycosides	Chuckles et al. 2003
22.	*Cucumaria frondosa*	Frondogenin	Findlay and Anand 1984.
23.	*Holothuria scabra.*	triterpene glycosides	Han et al., 2009

6.4 ANTI-INFLAMMATORY ANTI-BACTERIAL, ANTI-MICROBIAL, ANTI-FUNGAL, ANTI-YEAST AND ACTIVITIES

One of the earliest report on the anti-inflammatory property of sea cucumber meat was that of Whithouse and Fairlie (1994). They studied the effect of a human food supplement (SeaCare) composed of dried extracts from specific varieties of holothurians (sea cucumbers) and a sea plant and found to have anti-inflammatory activity in both sexes of two strains of rats. It is slightly less active than aspirin (w/w) against the acute carrageenan-induced paw inflammation, but without the gastrotoxicity of aspirin. It was also active against adjuvant-induced polyarthritis in rats on a daily dose schedule. Hydrocortisone is the widely the used wonderful anti-inflammatory agent and pain reliever today, but it can produce serious side effects. Notably, hydrocortisone can raise blood pressure, increase cholesterol, weaken bones, thin the skin, and depress immune function. Sea Cucumber extracts can reduce the pain and stiffness associated with arthritic disease, bursitis, and other problems of the bones and joints without any of the problems associated with hydrocortisone. Athletes also use Sea Cucumber to promote healing of cartilage injuries, which can occur as part of normal wear and tear of an active lifestyle. These properties are supported by number of investigators (Mena-Bueno *et al.*, 2016, Su X *et al.*, 2011).

6.5 REFERENCES

1. Alessandro Lovatelli, C. Conand, Food and Agriculture Organization of the United Nations. *Advances in sea cucumber aquaculture and management*: *Volume 463 of FAO fisheries technical paper* United Nations Food & Agriculture Org., 2004. ISBN 978-92-5-105163-4. 425 pages: 58.

2. Alexander.S.A., S.Avilov, Antoly I.K. , Stanislavsky D.V., Pavel S.D., Evgeny V.E.,Vladimir I K. Alexey V, S., SergiTaboada, Manuel B., Coxita Avila and Valentin A.S. Triterpene Glycosides from Antarctic she cucumbers. 1. Structure of Louvillosides A1, A2, A3, B1 and B2 from the sea cucumber *Staurocucumisliouvillet*. New Procedure for Separation of High Polymer Glycoside fractions and taxonomic revision. Journal of Natural Products October 8, 2015' vol.71, p.1677

3. Ali Reza Salarzadeh, Majid Afkhami, Darvish Bastami, Maryam, Chsanpour, Aida, Khazaal and, Mokheci. Proximate Composition of two species *Holothuriaarenicola* in Persian Gulf. Annals of Biological Rsearch 2012,3(3),1305-1311.

4. Althunibat OY[1], Ridzwan BH, Taher M, Daud JM, Jauhari Arieflchwan S, Qaralleh H. Antioxidant and cytotoxic properties of two sea cucumbers, Holothuriaedulis lesson and StichopushorrensSelenka. ActaBiol Hung. 2013 Mar;64(1):10-20.

5. Attoub S[1], Arafat K, Gélaude A, Al Sultan MA, Bracke M, Collin P, Takahashi T, Adrian TE, De WeverO.Frondoside a suppressive effects on lung cancer survival, tumor growth, angiogenesis, invasion, and metastasis.PLoS One. 2013;8(1):e53087. doi: 10.1371/journal.pone.0053087. Epub 2013 Jan 8.

6. Avilov S. A., V. I. Kalinin and V.A. Stonik., Three uncommon mono-sulphatedtriterpene glycosides, mollisosides A(2), B1(3) and B3(4) from sea cucumber *Astalostichopusmollis* , J. Natural Products. May 17, 2005, 6, 842-847.

7. Bakus, G.J. Defensive mechanisms and ecology of some tropical holothurians. Mar. Biol. 1968, 2, 23–32.

8. Bialy Z., M.Jurzyzsta, M. Melba and A. Tawe. Triterpenesaponins from aerial parts of *Medicagoarabica*. L. Journal of Agricultural and Food Chemistry, 2004,52:1095-1099.

9. Bordbar, S.; Anwar, F.; Saari, N. High-value components and bioactives from sea cucumbers for functional foods. A review. Mar. Drugs2011, 9, 1761–1805.

10. Brusca, R.C.; Brusca, G.J. Invertebrates, 2nd ed.; Sinauer Associates, Inc.: Sunderland, MA, USA, 2003; p. 936.

11. Chang -Lee M. V., R. J. Price and Lampilla. Journal of Food Science. 2006, 12, 344-353.

12. Chuckles H.D., A.P.Murray, A.M.Selda and M.S.Mair. Biologically active triterpene glycosides from sea cucumbers (*Holothurioidea, Echinodermata)*. In. Rahman A. (Ed.). Studies in Natural Products Chemistry,2003,28: 587-616' Elsevier Science, Amsterdam.

13. Chudil H. D. Muniain C.C., Seides A. M. and Maimer M.S. Cytotoxic and Antifungal triterpene glycosides from Patagonia sea cucumber *Hemoeiedemaspectablis*. J. Nat. Products. 2002, 65, 860-865.

14. Collin, Peter David, Sea cucumber carotenoid lipid fraction products and methods of use. US Patent: PCT. No. PCT/US99/01179. PCT. Pub.No. 4099/37314 PCT. Pub. Date July 29, 1999.

15. Conand C. Sea Cucumber Biology, Taxonomy, Distribution: Conversation Status. Proceeding of the Convention on International Trade in Endangered Species of Wild Fauna and Flora Tech Workshop on the Conversation of Sea Cucumbers in the Families Holothuridae and Stichopodidae; Kuala Lumpur, Malaysia. 1–3 March 2004.

16. Dmitry L. Aminin, Ekaterina S. Menchinskaya, Evgeny A. Pisliagin, Alexandra S. Silchenko, Sergey A. Avilov and Vladimir I. Kalinin , Anticancer Activity of Sea Cucumber Triterpene Glycosides *Mar. Drugs* 2015, 13(3), 1202-1223; doi:10.3390/md13031202.

17. Dyck, S.V.; Caulier, G.; Todesco, M.; Gerbaux, P.; Fournier, I.; Wisztorski, M.; Flammang, P. The triterpene glycosides of Holothuriaforskali: Usefulness and efficiency as a chemical defense mechanism against predatory fish. J. Exp. Biol. 2011, 214, 1347–1356.

18. Dyck, S.V.; Gerbaux, P.; Flammang, P. Qualitative and quantitative saponin contents in five sea cucumbers from the Indian Ocean. Mar. Drugs 2010, 8, 173–189.

19. Fell HB. Phylum Echinodermata. In: Marshall AJ, Williams WS, editors. Text Book of Zoology Invertebrates. 7th ed. American Elservier; New York, NY, USA: 1972.

20. Findlay. J. A. and AnandDaljeet. Frondogenin-A, new Aglycone from sea cucumber, *Cucumariafrondosa*, J. Nat. Products, March- April,1984. vol. 47., No. 2, pp. 320-324.

21. Flammang, P.; Ribesse, J.; Jangoux, M. Biomechanics of adhesion in sea cucumber Cuvierian tubules (Echinodermata, Holothuroidea). Integr. Comp. Biol. 2002, 42, 1107–1115.

22. G. Moraes, PT. Northcole, A.S.Antonov, A.I. Kalinovsky, Pavel S., S.A. Avilov, V.I. Kalinin and V. A. Stonik. Mollisosides A, B1 and B2' Minor Triterpene glycosides from NewZealand and South Australian Seacucumber (*Australostichopusmollis*). J Natural Products. May17, 2005, pp.842-847.

23. Han H, Yi Y, Xu Q, La M, Zhang H. Two new cytotoxic triterpene glycosides from the sea cucumber *Holothuriascabra*. Planta Med. 2009; 75:1608–1612. [PubMed].

24. Hong Zhong, Mohammad Ahamad Khan and FerridonShahidi.Compositional Characteristics and anti oxidant properties of Fresh and Processed sea cucumber (*Cucumariafrondosa*). Journal of Agricultural and Food Chemistry. 2007, vol. 55: 4. 1186-119.

25. HuiXu, Jingfeng Wang, Yaoguang Chang, JieXu, Yuming Wang, Tenteng and Changhu Xue., Fucoidan from the sea cucumber *Acaudinamolpadioides* exhibits ant-adipogenic activity by modulating the WNT/ Beta-Catenin pathway and down regulating the SREMP-Ic expression. Food Funct. 2014, 5: 1547-1555.

26. Ignacio Sottorff, AmbbarAbally, Victor Hernandez, Louis Roa, Lilian X. Munoz, Moria Silva, Jose Becerra and Allisson Astuya. Resistance de Biologiay Ocenogratia.,April 2013, Vol. 48., No. 1.: 23-35.

27. Ismail H, M. Jurists, M.Mellaand A.Tawa.,Triterpenesaponins from aerial parts of Medicagoarabica. Journal of Agricultural and Food Chemistry.2004,52: 1095-1099.

28. Jean Mamelona, EmillenPelletier, Kari Girard- Lalaneetta, Jean Legault, SalwaKarboune and SelimKemasha. Quatification of phenolic contents and antioxidant capacity of Atlantic sea cucumber (*Cucumariafrondosa*). Food Chemistry. 2007, Vol. 104(3): 1040-1047.

29. Jun Wu, Yang-Hua Yu, Hai-Feng Tang, Hou-Ming Wu, Zhen-Rong-Zou, Hou-Wen Lin. Noblisides A-C, Three New Triterpene Glycosides from sea cucumber *Holothurianobilis*. Planta Med, 2006; 72(10): 932-935.

30. Kalinin V.I., N.G.Prokofieva, G.N. Likhatskaya, E.B. Schentsova, I.G. Agafonova, S.A. Avilov, and O.A. Drozdova, Hemolytic activities of triterpene glycosides from the holothurian order *Dendrochirotida*. Some trends in the evolution of this group of toxins,Toxicon ,1996, 34: 475-483.

31. Kim SK[1]andHimaya SW. Triterpene glycosides from sea cucumbers and their biological activities.Adv Food Nutr Res.2012;65:297-319. doi: 10.1016/ B978-0-12-416003-3.00020-2.

32. Kobayashi M., M.Hori, K. Khan, T. Yasuzawa, M.Matsui, S. H. Suzuki and I. Kithara, Marine Natural Products. XXVII. Distribution of lanostane-type triterpeneoligoglycosides in ten kinds of Okinawan Sea cucumber. Chemical and pharmaceutical Bulletin, 1999, 49: 726-731.

33. M.V. Chang- Lee, R, J. Price, L. E. Lampilla. Proximate Composition of Sea cucumbers. Journal of Food Science. 2006,12,344-353.

34. Masa fume Kaneko, Fumiaki Kisa, Koji Yamada, Tomofumi Miyamoto and Ruichi Higuchi. Structure of Neuritogenic Active Ganglioside from seacucumber (*Stichopusjaponicus*). European Journal of Organic Chemistry. 1999 November vol. 11, pp.3171-3174.

35. Mena-Bueno S, Atanasova M, Fernández-Trasancos Á, Paradela-Dobarro B, Bravo SB, Álvarez E, Fernández ÁL, Carrera I, González-Juanatey JRandEirasS.Sea cucumbers with an anti-inflammatory effect on endothelial cells and subcutaneous but not on epicardial adipose tissue.FoodFunct. 2016 Feb;7(2):953-63. doi: 10.1039/c5fo01246e.

36. Nigrelli, R.F. The effects of holothurin on fish and mice with sarcoma 180. Zoologica (NY)1952, 37, 89–90.

37. Ping Dong, Chang-huXue, Lin-fang, JieXu and Shi-guo Chen. Determination of Triterpene Glycosides in Sea cucumber (*Stichopus japonicas*) and its related products by High Performance Liquid Chromatography. Journal of Agricultural and Food Chemistry., 2008, Vol. June 17, pp. 4937-4942.

38. Ridzwan BH. Sea Cucumbers, A Malaysian Heritage. 1st ed. Research Centre of International Islamic University Malaysia (IIUM); Kuala Lumpur Wilayah Persekutuan, Malaysia: 2007. pp. 1–15..

39. Rifkin, J.F. Venomous and Poisonous Marine Animals: A Medical and Biological Handbook; Burnett, J.W., Fenner, P.J., Eds.; UNSW Press: Kensington, NSW, Australia, 1996.

40. Rodriguez. J. R.,R. Castro and R. Riguera. Holothurinosides.new antitumor non-sulphatedtriterpenoid glycosides from sea cucumber, *Holothuriaforsakali*. Tetrahedron,1991, 47: 4753-4762.

41. Stonic V. A. and Elyatov G. B. secondary metabolites from echinoderms as chemotaxonomic markers, In. Scheur P. J. Editor Bioorganic Marine Chemistry, Berlin; Springer 1988: 43-88.

42. Su X, Xu C, Li Y, Gao X, Lou Y, *et al.* (2011) Antitumor Activity of Polysaccharides and Saponin Extracted from Sea Cucumber. J Clin Cell Immunol 2:105. doi:10.4172/2155-9899.1000105

43. Subhuti, Dharmananda, Sea cucumber: Food and Medicine*ww.itmonline.org/ arts/seacucumber.*

44. T. Zo, Z.Li., Y. KV., G.Duvan, C. Wang., Q. Tang and C. Xue. Rapid Identification of Sea cucumber species with Multiplex- PCR. Food Control. July 2012, vol. 26(1) 58-62.

45. TangWeici, *Chinese medicinal materials from the sea*, Abstracts of Chinese Medicine 1987; 1(4): 571-600.

46. Tian F., Zhang X., Tong Y., Yi Y., Zhang S., Li L., Sun P., Lin L., Ding J. PE, a new sulfated saponin from sea cucumber, exhibits anti-angiogenic and

anti-tumor activities *in vitro* and *in vivo*. Cancer Biol. Ther. 2005; 48: 874–882. doi: 10.4161/cbt.4.8.1917. [PubMed] [Cross Ref]

47. Tian F., Zhu C.H., Zhang X.W., Xie X., Xin X.L., Yi Y.H., Lin L.P., Geng M.Y., Ding J. Philinopside E, a new sulfated saponin from sea cucumber, blocks the interaction between kinase insert domain-containing receptor (KDR) and alphavbeta3 integrin via binding to the extracellular domain of KDR. Mol. Pharmacol. 2007; 72:545–552. doi: 10.1124/mol.107.036350. [PubMed] (Cross Ref).

48. Whitehouse MW and DP Fairlie. November 1994.Anti–Inflammatory activity of a holothurian (sea cucumber) food supplement in rats. Inflammopharmacology 2(4):411-417

49. Yang Peiying, *et al.*, Inhibition of proliferation of PC3 Cells by the branched-chain fatty acid, 12-MTA, is associated with inhibition of 5-lipoxygenase, The Prostate 2003; 55: 281-291

50. ZhangEnchin (Chief Editor), Chinese Medicated Diet, 1988 Publishing House of Shanghai College of Traditional Chinese Medicine, Shanghai.

51. Zhong Y, Khan MA, Shahidi F, Compositional characteristics and antioxidant properties of fresh and processed sea cucumber (Cucumariafrondosa).J Agric. Food Chem.2007 Feb 21;55(4):1188-92. Epub 2007 Jan 23.

CHAPTER **7**

ALKOXY GLYCEROLS

A lkoxyglycerols are components of the Diacylglycerol Ethers(DAGE). The diacylglycerols ethers are glycerides consisting of two fatty acids chains covalently bonded to a glycerol molecule through ester linkages, and their chemical structure consist simply of a backbone of glycerol attached with an ether link to an alky group.

Ethers of glycerol with long chain fatty alcohols were first discovered by Tsujimoto and Toyama (1922) from the non-saponfiable fraction of the liver oil of certain elasmobranches.

Alkoxy glycerol's (AKG) are a group of compounds composed of ether linked glycerol found in all living organisms. It is one of the principal components in milk of all mammals and also in bone marrow. AKGs are naturally occurring in the organs such as bone marrow, spleen liver, lymphatic tissues and blood (Hallgren B. and Larsson S., 1962). Their principal role is to synthesize blood and hence called hemotopoietic compounds. AKGs are found in human colostrums, milk of cow, sheep, red cells and blood plasma. Human milk contains 10 times more unsubstitued glycerol ethers than that of cow and sheep (Hallgren *et al.* 1974). The highest amount of alkylglycerol was found in the liver oil of Greenland sharks, the gray dog fish and rat fish (Hallgren B. and Larsson S., 1962) and Indian deep-seashark (Certophorus Spp.) caught from sea around Andaman and Nicobar Islands (Gopakumar, 1997)

Deep sea shark liver oil from Centrphorus spp. from Indian Ocean is a rich source of squalene and alkoxy glycerol (Gopakumar, 1997). Both squalene and alkyl glycerol are commercially produced now in State of Kerala, India by Asha Biochem at Vadakara and marketing globally. Following is the yield;

Table 16: Percentage of Glycerol Esters in Lipids

Source	Glycerol Esters in Lipids(as wt%)
Liver oils of Elasmobranches	10- 30
Human bone marrow	0.2
Human milk	0.1
Human spleen	0.05
Cow's milk	0.01
Shark liver oil (Centrphorus spp.)*	5-!0%

(**Source**: Hallgren B and Larsson S. The glyceryl esters in man and Cow. J. Lipid Res. 3(1): 39-43, 1962)

* Data Courtesy, Asha Biocem, Vdakara, Kerala, India

Table 17:   Yield of Alkoxy glycerol from liver oil from deep sea shark (Centrophorus Spp.) in Indian Ocean

Centrophorus Spp.	Yield
Squalene	70-80% by wt. of liver oil
Di acyl glycerol ether (DAGE)	68-70% of 20% mother oil (after squalene extraction)
Alkoxy Glycerol	20-22 % of DAGE

Data ; Courtesy, Asaha Biochem Pvt. Ltd. Vadakara, Kerala State , India

7.1 COMPOSITION OF ALKOXY GLYCEROLS

The principal components of AKGs are high molecular alcohols (saturated and unsaturated), fatty acids and glycerol. These high molecular alcohols exist in nature as their Di acyl glycerol ether form. Some important alcohols seen in AKGs are given below in table 3. Naturally occurring Alkoxyglycerols are mixtures with varying number of carbon atoms in the side chain varying chain lengths from 14-24. Chimyl alcohol and Batyl alcohol are saturated glycerol ethers having 16 and 18 carbon atoms. respectively in the side chain. Selachyl alcohol is an unsaturated glycerol ether with 18 carbon atoms (Table 18). Methoxy substituted glycerol ethers are also occurring in nature. In these compounds one hydrogen atom, mostly on carbon 2 is substituted by a methoxy group($-0CH_3$).

The presence of glyceryl ethers in land animals was shown for the first time by Holmes et al. (1941). Hallgren and Larsson (1962) found that in human spleen chimyl, batyl and selachyl alcohols constituted about 85% of glyceryl ethers and in human marrow, the saturated ethers65% of total ethers. Subsequently several workers have described the presence of large amounts of alkylglycerols in deep sea shark liver oil and its beneficial effects and applications in chemotherapy. (Pugliese *et al.*, 2008. Beniamino *et al.*, 2014, Anne-Laure Deniau *et al.*, 2010).

Table 18: Important alcohols naturally occurring in Alkoxy glycerols

No,	Name	Molecular weight	Melting Point, C°	Chemical structure
1	Chimyl alcohol	$C_{19}H_{40}O_3$	60-61 (Solid)	$CH_3\text{-}(CH_2)_{15}\text{-}O\text{-}CH_2\text{-}CHOH\text{-}CH_2OH$
2	Baty alcohol	$C_{21}H_{44}O_3$	70-71 (Solid)	$CH_3\text{-}(CH_2)_{17}\text{-}O\text{-}CH_2\text{-}CHOH\text{-}CH_2OH$
2	Selachyl alcohol	$C_{21}H_{42}O_3$	Liquid	$CH_3\text{-}(CH_2)_7\text{-}CH=CH\text{-}(CH_2)_8\text{-}O\text{-}CH_2\text{-}CHOH\text{-}CH_2OH$

Biosynthesis of Alkoxy Glycerols

Di acyl Glycerol ethers are hydrolyzed by intestinal enzymes to their free ether form called alkoxy glycerols.

Hydrolysis of Diacyl Glycerol Ethers by enzymes

$CH_2 - O - R$ (OH–) $CH_2 - O - R$
| |
$CH - OOCR_1$ $\longrightarrow$ $CH\text{-}OH + R_1COOH + R_2COOH$
| |
$CH_2 - OOCR_2$ **Hydrolysis** CH_2OH **Fatty acids**

Diacyl Glycerol Ether   **Alkoxy Glycerol**

7.2 SHARK LIVER OIL AND ALKYLGLYCEROL

Shark liver oil has been used by fishermen and coastal people around the world as a remedy for tiredness and general debility. The discovery of the presence of Vitamin D and E in shark liver oil, one of the richest source, has further enhanced the therapeutic importance of shark liver oil. Historically shark liver oil from deep sea shark was used by fishermen in Norway and Sweden in Europe and Japan for several specific applications like wound healing, removal of tiredness and treatment of irritation of respiratory and alimentary tracks. In India it was used by traditional Aurvedic physicians for treatment of burns and several skin diseases. The discovery in the 20th century that shark liver oil is one of the richest source of several biologically active molecules like squalene, pristine, esters of fatty acids, glycerol ethers, triglycerides, cholesterol, fatty acids. Vitamin A, D and E has further widened its clinical uses to mankind. The ratfish (Chimera*monstrosa),* belonging to the family of elasmobranch class like the dogfish, liver oil is also a rich source for alkylglycerols. It was used to treat a glandular disease in ancient days. Today this disease is called lymphadenopathy (Melvyn R Werbach 1994.). The composition of shark liver oil varies from species to species and depends on the size of the shark feeding habits, gender, growth rate, depth at which it lives, seasons and ocean temperature. On an average the liver constitutes about 25% of the total shark body weight.

Greenland shark liver oil contains 1-O-(2-hydroxyalkyl) glycerols, 1-O-(2-methoxyhexadecyl)-glycerol, 1-O-(2-methoxy-4-hexadecenyl)-glycerol, and 1-O-(2-methoxy-4-octadecenyl)-glycerol (Hallgren and Ställberg G 1967, 1974). There is now substantial experimental evidence to show that the alkylglycerols are immune stimulants. Glycerol esters isolated from Greenland shark liver oil administered orally to mice was shown to stimulate immune reactivity. The evidence presented suggests that this effect was due specifically to the methoxy-substituted glycerol'sethers. (Melvyn R. W, 1994).

Alkyl Glycerol(DDGE/AKG) is now manufactured and marketed in India by leading Pharmaceutical company and their product qualities are given below.Specifications of alkylglycerol from *Centrophorus spp.* of Indian Ocean species are provided below (Table 19).

Table 19: Specification of alky glycerol

PRODUCT SPECIFICATION

Product: Shark liver oil containing DAGE/AKG

PROPERTIES		LIMIT
1.	Colour and appearance	:	Golden yellow liquid
2.	Specific gravity	:	0.902 to 0.910
3.	Acid value	:	Less than 2 mg KOH per gm
4.	Iodine value	:	95 to 110
5.	Saponification value	:	140 to 145
6.	Unsaponifiable matter	:	25 to 30%
7.	Peroxide value	:	Less than 5 meq/kg
8.	Alkoxy glycerol content	:	22 to 25 %
9.	Heavy metals	:	Within the allowed limits.

Method of Analysis

1. Indian Pharmacopia; 2. AOAC Official Methods of Analysis 1990.

Data courtesy, Aasha Biochem, Chorode, Vatakara, Kerala – 673106. Drug Manufacturing License No: 21/28/2003, Drug Controller, Kerala, India, Dated 4/6/2003.

7.2.1 Role of AKGs in in Immune response

Alkoxyglycerols are natural substances found in bone marrow and in mother's milk. They have the ability to enhance the immune system by helping produce higher quantities of white blood cells and macrophages which are the first line fighters against foreign matter or organisms in our body. The Alkoxyglycerols found in mother's breast milk is a major factor in providing immune support to the infant. It provides infants with natural protection and immunity against infection as well as helping the continued development of their immune system. Science has proven that breast fed babies are more resistant to infection throughout their lives because of the more abundant ingestion of Alkoxyglycerols in their early life. Human milk and colostrums have been shown to possess a wide variety of immune stimulating properties by a number of researchers (Hanson L.A. and Winberg J,1972, Orga *et al.* 1977, Prentice A. 1987). This property is due to the presence of Alkylglycerols in human milk. The Alkoxy glycerol were found to stimulate production of lekcocytes and thrombocytes and activation of macrophage and antitumor activity (Suk Y. OH and Lalitha S. J. 1994).It is now well established that Alkoxyglycerols enhance the immune system through promoting bone marrow proliferation but also that Alkoxyglycerols enhance the efficiency of the immune system by directly increasing the phagocytic function of white blood cells called macrophages. Later it was verified that Alkoxyglycerols help protect the human body from a variety of radiation injuries by enhancing immune function (Brohult A. 1977). Alkylglycerols are showing multiple biological activities One theory proposed is that they may be incorporated into cell membrane phospholipids and from there modify their physical properties like membrane fluidity antioxidant activities or alter signaling through phospholipase pathways (Debouy *et al.*, 2008). Alkylglycerols are thought to be a very promising principle to facilitate transport of therapeutics across the blood- barrier (Anadon *et al.,* 2010).

7.3 ALKOXYGLYCEROLS AND THEIR USE IN CANCER TREATMENT

Alkylglycerol activates macrophages. Macrophages are a type of white blood cells that engulfs and digests cellular debris, foreign substances, microbes, cancer cells, and anything else that does not have the types of proteins specific of healthy body cells on its surface in a process called phagocytosis. Alkyl glycerols are inflammation products of lipids in cancer tissues and are potent macrophage stimulating agents.

Neoplastic tissues have high levels of alkylglycerol esters, together with characteristic abnormalities of ether-lipid synthesizing and degrading enzymes suggesting that alkylglycerol administration may somehow impede cellular metabolism. (Snyder, 1972) The Brohult *et al.* (1979) have reported that the alkylglycerols or their esters form liquid crystals rigidify the cell membrane, thus reducing the cell's ability to divide. The alkylglycerol is naturally occurring compounds in animal bodies and their level rises within tumor cells as an attempt to control cell growth. One of the essential step in cell proliferation involves activation of protein kinase C and that can be inhibited by alkylglycerols (Skopinska *et al.,* 1999). Shark liver oil demonstrated inhibitory actions against cutaneous angiogenesis in certain cancer cells in mice and that property is due to the presence of alkylglycerols.

Dodecylglycerol (DDG) is one major alkylglycerol that is found to initiate macrophage activation processes by releasing and transmitting signal factors to T-cells and which in turn modified to produce factors stimulating macrophages for ingestion capability (Yamamoto *et al.* 1988). AG was found to inhibit infectious HIV—1 production and induce defective virus formation (Louis *et al.,* 1990). The biological effects of alkylglycerol on cancer treatment has been extensively studied by several workers in the past (Edlund 1974, Linman1960, Osmond *et al.,* 1963, Werbach 1994). The therapeutic effect of alkylglycerols was first demonstrated to the world by a Swedish doctor Astrid Brohult in 1952 at the Radiumhemmet Hospital in Stockholm, Sweden, where she was working. She did some study by supplying bone marrow extract from calf bones to leukemic children with leucopenia and found that it improved the white blood cell formation. But it was her husband Sven Brohult, a Professor of Biochemistry at a Swedish university, after years of research established the fact that the immune stimulants were alkylglycerols and they have excellent anticancer properties. (Brohult *et al.,* 1970, 1977, 1978,1979, 1986,). Later Anadon *et al.,* (2010) did extensive study by supplying alkoxyglycerol in diets of rats and proved that there are no changes in hematological and serum chemistry values, organ weights, or gross or histological characteristics showing that alkylglycerol has no toxic effect and safe to consume.

Investigations carried out in the Department of Oncology and Haematology in Odense Hospital in Denmark (.Hasle H and Rose C, 1991) revealed that there is no documented evidence to support that treatment with Alkoxyglycerol inhibited tumour growth or reduced mortality. The number of cases of irradiation damages were found to be fewer in the groups treated with alkoxyglycerol, but the difference may be partially explained by different subdivision into stages. Alkoxyglycerol results in increase in the leukocyte and thrombocyte counts while higher or lower doses have, apparently, the opposite effect. The available literature concerning the clinical effect of alkoxyglycerol is limited and unsystematic and does not support the employment of alkocyglycerol in the treatment of cancer.

7.3.1 Recommended Daily Dosages

As an adjunct therapy to traditional cancer patients for prophylactic and therapeutic measures in immunity deficiency the maximum recommended usage is 600mg alkoxyglycerols per day (*i.e.,* 10mgkg^{-1}). In humans, dosages of 100mg three times a day have not shown to cause adverse effects. (Pugliese e.t al.,1998).

7.4 REFERENCES

1. Anadon A., Maria A, Irma A, Eva R, Francisco J., Guillermo R., and Carlos T. Acute and Repeated dose(28days) oral safety studies of an alkoxyglycerol extract from shark liver oil in rats. J. Agric. Food Chem. 2010,58, 2040-2046

2. Anne-Laure Deniau, Paul Mosset, Frederique Pedrone, romain Mitre Damien Le Bot and Alain B Legrand. Multiple Beneficial Health effects of Natural Alkylglycerols from shark Liver Oil. Mar. Drugs 2010, 8, 2175-2184.doi: 10.3390/md 8072175.

3. Beniamino P., Alfonso P and Alessandro Di Cerbo. Jurassic surgery and immunity enhancement by alkylglycerols of shark liver oil. Lipids in Health and disease 2014 13 : 178

4. Berdel WE. Ether Lipids and analogs in experimental cancer therapy. A brief Review of the Munich experience. Lipids 1987; 22: 970-973.

5. Brohult A, Brohult J. and Brohult S. and Joelsson I Reduced mortality in cancer patients after administration of alkylglycerols. 1986, Acta Obstet Gynecol Scand 65: 779-85.

6. Brohult A, Brohult J. and Brohult S. and Joelsson I. Effect of alkoxyglycerols on the frequency fistulas following radiation therapy of carcinoma of the uterine cervix. Acta Obstet Gynecol Scand. 1979. 58(2): 203-7.

7. Brohult A, Brohult J. and Brohult S. and Joelsson I. effect of alkoxyglycerols on the frequency of injuries following radiation therapy for carcinoma of the uterine cervix. Acta Obstet Gynecol Scand 1977. 56; 441-8.

8. Brohult A, Brohult J. and Brohult S. Biochemical effects of alckoxyglycerols and their use in cancer therapy. Acta Chem Scand 1970, 24: 730.

9. Brohult A, Brohult J. and Brohult S. Regression of tumor growth after administration of alkoxyglycerols. Acta Obstet Gynecol Scand 1978. 57(1); 79-83.

10. Brohult A. Alkoxyglycerol-esters in irradiation treatment. Advances in Radiobiology: Proceedings of the Fifth International Conference in Radiobiology, Stockholm, August, 1956. 1957, pp. 241-7.

11. Debouzy, J-C; crouzier, D.; Lefebve, B and Dabouis, V. study of alkoxyglycerol containing shark liver oil: a physic chemical support for biological effects. Drug Target Insights 2008,3, 125-135.

12. Edlund T. Protective effect of d,l—octadecylglycerol etherin micebody x-irradiation. Nature, 1974; 174:1102

13. Gopakumar. K Tropical Fishery Products. Oxford and IBH Publishing Company PVT LTD New Delhi. 1997.

14. Hallgren B and Ställberg G. 1-O-(2-hydroxyalkyl) glycerols isolated from Greenland shark liver oil. Acta Chem Scand B . 1974;28(9):1074-1076.

15. Hallgren B and Ställberg G. Methoxy-substituted glycerol ethers isolated from Greenland shark liver oil. Acta Chem Scand . 1967;21(6):1519-1529.

16. Hallgren B. and Larsson S. 1962 The glyceryl ethers in man and cow. J. Lipid Res. 3: 39-43.

17. Hallgren B. and Larsson S. 1962. The glyceryl ethers in the liver of oils of elasmobranch fish. J. Lipid Res. 3:31-38.

18. Hallgren B., Niclasson A., Stalberg G. and Thorin H. 1974 On the Occurrence of 1-O- alkylglycerols and 1-O-(2 methoxy alkyl) glycerols in human colostrums, humam milk, cow's milk, sheep's milk, human red bone marrow, red cells, blood plasma and a uterine carcinoma. Acta. Chem. Scand B28:1029-1034.

19. Hansen L. a. and Winberg J. 1972 Breastmilk and defence against infection in newborns. Arch Dis. Child. 47: 845-848.

20. Hasle H, Rose C.Shark liver oil (alkoxyglycerol) and cancer treatment. Ugeskr Laeger. 1991 Jan 28; 153(5):343.

21. Linman JW. Hemopoeitic effects of glycerol ethers. III. Inactivity of selachyl alcohol. Proc. Soc. Exp. Biol. Med. 1960;104-: 703-6.

22. Louis S. Kucera, Nathan Iyer, Eva Leake, Adam Raben, Edward j modest, Larry W. Daniel and Claude Piantadosi. Novel Membrane- Interactive ether Lipid Analogs that Inhibit infectous HIV-1 Production and induce Defective Virus Formation. AIDs Research and Human R etroviruses. April 1990,6(4); 491-501.

23. Orga SS. Weintraub D, Orga PL. 1977 Immunologic aspects of human colostrums and milk. J. Immunol. 119: 245-248.

24. Osmond DG, RoylanceP., Webb AJ. And Yoffey JM. The action of batyl alcohol and selachyl alcohol on the bone marrow of of the guinea pig. Acta Hematol 1963;29; 180-6.

25. Prentice A. 1987.Breast feeding increases concentrations of IgA in infants' urine. Arch. Dis. Child 62: 792-795.

26. Pugliese, P.T., KarinJordan, Hokan Cederberg and Johan Brohult. Some Biological Actions of Alkylglycerols from shark liver oil. J. of Alternative and Complimentary Medicine 2008, 4(1): 87-99.

27. Skopinska-Rózewska E, Krotkiewski M, Sommer E, *et al.* Inhibitory effect of shark liver oil on cutaneous angiogenesis induced in Balb/c mice by syngeneic sarcoma L-1, human urinary bladder and human kidney tumour cells. Oncol Rep. 1999;6(6):1341-1344.

28. Snyder F. Ether Lipids. Chemistry and Biology. New York, Academic Press, 1972.

29. Suk Y. OH and Lalitha S. Jadhav.1994 Effect of Dietary Alkylglycerols in Lactating Rats on Immune Response in Pups. Pediatric Research 36, No.3, 300-305.

30. Tsujimoto, M. and Toyama, Y., 1922. Uber die unverse if baren Bestandteile (hoheren Alkohole) der Haifish und Rochen-leberole.I. Chemische Umschau 29:27-29.

31. Werbach MR. Alkylglycerols and Cancer, J. Orthmol Med. 1994:9.95—101.

32. Werbach MR. Alkyl Glycerols and Cancer. J. Orthomol.Med. 1994; 9(2), 95-102.

33. Yamamoto N., St Claire DA Jr., Homma S and Ngwenya BZ. Cancer Research 1988, Nov. 1; 48(21);6044-9.

CHAPTER 8

MUSSELS

Mussel is the common name used for members of several families of bivalvemolluscs, from saltwater and freshwaterhabitats. These groups have in common a shell whose outline is elongated and asymmetrical compared with other edible clams, which are often more or less rounded or oval. The word "mussel" is most frequently used to mean the edible bivalves of the marine family Mytilidae most of which live on exposed shores in the intertidal zone, attached by means of their strong byssal threads ("beard") to a firm substrate (A few species (in the genus *Bathymodiolus*) have colonised hydrothermal vents associated with deep ocean ridges.

Fig. 35: Green Mussel (Perna virridis)

The mussels are bivalves eaten by man for thousands of years and is a food delicasy. The most commonly eaten mussel is the marine species called mytilidae. But there are 17 wellknown edible species. Mussels belong to the Kingdom of Animalia, phylum Mollusca and class Bivalvia. There are threesub-classes :

● Marine mussels (Pteromorphia)

● Freshwater mussels (Palaeoheterodonta)

● Zebre mussels (Heterodonta)

Proximate composition : Freeze Dried Mussel (*Perna virridis*)

Mussel meat is a rich source of vitamin A, B-vitamins like folic acid, B12 and minerals like phosphorus, zinc and manganese as well as omega-3 fatty acids (Table 20, 21, 22, 23, 24).

● Moisture : 4.5%

● Protein: 48.0%

● Ash: 12.4%

● Fat: 5.9%

Table 20: Minerals and Heavy Metals Composition

	Cu ppm	Zn Ppm	Mn ppm	Mg ppm	Co ppm	Fe ppm	Cd ppm	Kg%	Nag%	Cag%
FD Mussel	18.5	49.5	6.9	3594.9	ND	361.13	ND	0.85	2.21	2.23

Table 21: Amino acid composition

S. No.	Amino acid	FD Mussel (g/16gN)
1.	Aspartic acid	10.3
2.	Threonine	5.3
3.	Serine	5.3
4.	Glutamic acid	7.9
5.	Proline	5.1
6.	Glycine	25.7
7.	Alanine	6.4

[Table Contd.

Contd. Table]

S. No.	Amino acid	FD Mussel (g/16gN)
8.	Cysteine	–
9.	Valine	2.9
10.	Methionine	1.7
11.	Isoleucine	4.7
12.	Leucine	6.8
13.	Tyrosine	2.1
14.	Phenyl alanine	4.6
15.	Histidine	–
16.	Lysine	7.1
17.	Arginine	8.1

Table 22: Fatty acid composition (Freeze Dried Mussel) (Major fatty acids)

Fatty acid	% of total fatty acids
C14	17.8
C16	5.8
C18	18.6
C18:1	6.3
C18:3	8.9
C20:1	12.1
C20:4	18.3
C20:5	1.0
C22:6	5.2

Table 23: PAH Analysis

1.	Naphthalene	Not detected
2.	Acenaphthylene	Not detected
3.	Acenaphthene	Not detected
4.	Fluorene	Not detected
5.	Phenanthrene	Not detected
6.	Anthracene	Not detected
7.	Fluoranthene	Not detected

[Table Contd.

Contd. Table]

8.	Pyrene	Not detected
9.	Benzo (a) anthracene	Not detected
10.	Chrycene	Not detected
11.	Benzo (b) fluoranthene	Not detected
12.	Benzo (k) floranthene	Not detected
13.	Benzo (a) pyrene	Not detected
14.	Dibenzo (a,h) anthracene	Not detected
15.	Benzo (g,h,i) perylene	Not detected
16.	Indeno (1,2,3-CD) pyrene	Not detected
17.	Geosmin	Not detected

Table 24: Pesticides

Component	
α - BHC	Not detected
β- BHC	Not detected
γ - BHC	Not detected
Heptachlor	Not detected
Aldrin	Not detected
Heptachlor Epoxide Isomer B	Not detected
pp DDE	Not detected
Dieldrin	Not detected
op DDD	Not detected
Endrin	Not detected
pp DDD	Not detected
op DDT	Not detected
pp DDT	Not detected

8.1 GREEN –LIPPED MUSSEL (*Perna canaliculus*)

New Zealand's Green – Lipped Mussel is now widely used to treat joint pains.Developed by leading Marine Scientist John Croft at and currently manufactured and marketed by Vitaco Health (NZ) Ltd. is widely used world over under trade name "Seatoone". It was earlier marketed by a company formerly known as Healtheries of New Zealand Ltd.

Fig. 36: Green-lipped Mussel

8.2 MEDICAL APPLICATIONS OF MUSSEL MEAT

Mussel meat contains as a natural component substance called glycosaminoglycans. They belong to the group of polysaccharides called mucopolysaccharides. Glycosaminoglycans play a lead role in the production and repair of skin, tendons, synovial fluids and cartilage.In addition to that mussel meat also contain two compounds of importance namely glucosamine and chondroitin sulphates. Glocosamine is the building block in the biosynthesis of chondroitin sulphates in animals. Mucopoly saccharide molecules combines with proteins to form protoglycans. Protoglycans are highly hydrophilic in nature. They are the natural filling materials in all gaps, as filling materials in body joints. Binding with water they form very large slippery space filling molecules and act as lubricants and shock absorbers of joints. When this fluid dries joint pain starts and develop osteoarthritis, inability to move and eventually needing knee replacement.

8.2.1 Role of arachidonic acid in inflammation

Arachidonic acid produces three compounds by enzymatic conversion. They are prostaglandins (induce bleeding), thromboxanes and leucotrienes, the last two proinflammatory. The one that produces prostaglandins are produced by an enzyme called cyclo-oxygenase-1 (COX1). The proinflammatory compounds are produced by another enzyme called cyclo-oxygenase-2 (COX2). Normally joint inflammation is treated by giving steroids (Cortisone) and NASIDs (non- steroidal anti-inflammatory drugs like aspirin or iboprufin. But thy carry high risk of gastro-intestinal ulceration and bleeding. Hence cannot be used for long term treatments. But mussel meat is a basic food eaten by man for thousands of years. Studies conducted byMiller and Ormrod (1980) at Auckland Hospital in New Zealand have shown that long term use of Green- lipped mussel (Perna *canaliculus*) extract have shown no adverse side effects even consumed at 120 times of standard dosage.

It is now well established by clinical studies that Green-lipped -mussel extract selectively inhibits COX2 without affecting COX1(Jon Croft 2003.). It was also shown recently that Indian Green mussel Perna virridis also exhibit this property and Central Marine Fisheries Research Institute Kochi has prepared an extract in powder form and marketing it as a neutrceutical to treat osteoarthritis. It is likely that all green mussel species extracts may show this property and this needs future study in this field.

8.3 HEALTH BENEFITS OF MUSSEL MEAT

- As a slimming agent
- Prevent Hair Loss
- Regrow Thinning Hair
- Anxiety Depression Treatment
- Top Bodybuilding Supplements
- Best Memory Boosters
- Joint Supplements
- Immune System Boosters
- Benefits of Bone Broth
- Cancer Fighting Foods
- For the treatment of osteo-arthritis
- As anti-inflammatory agent
- Enzyme (COX2) inhibitor

8.4 REFERENCES

1. John Croft. 2003. Health from the Seas. *Freedom From Disease*: Vital Health Publishing.P. O. Box 152Ridgefield, Ct 0687.

2. Miller, T. E.and Ormord, DJ .1980. The anti-inflammatory activity of Perna canaliculus (New Zealand Green-lipped Mussel).NZ.Med. J. 667: 187-193

3. Whitehouse MW, Fairlie DP.1994, Anti-inflammatory activity of a holothurian (sea cucumber) food supplement in rats. Inflammopharmacology. 2:411-417.

4. *The Editors of Encyclopædia Britannica (22 May 2009). "Mussel". Encyclopædia Britannica.*

CHAPTER **9**

SEA HORSES

S eahorse is the name given to 54 species of small marine fishes in the genus Hippocampus. "Hippocampus" comes from the Ancient Greek word hippos (5ððïò, híppos) meaning "horse" and kampos (êÜìðïò, kámpos) meaning "sea monster"(Duvernoy,2005). The word "seahorse" can also be written as two separate words (sea horse), or hyphenated (sea-horse, Oxford Dictionary, 2007) having a head and neck suggestive of a horse.

Seahorses are fish. They live in water, breathe through gills and have a swim bladder. However, they do not have caudal fins and have a long snake-like tail. They also have a neck and a snout that points down. Seahorses have excellent eyesight and their eyes are able to work independently on either side of their head. This means they can look forwards and backwards at the same time! This is particularly useful as they hunt for food by sight.

Unlike most other fish, seahorses have an exo-skeleton. Their bodies are made up of hard, external, bony plates that are fused together with a fleshy covering. They do not have scales. Seahorses are poor swimmers. They rely on their dorsal fin beating at 30-70 times per second to propel it along. Pectoral fins either side of the head help with stability and steering. Seahorses can change colour very quickly and match any surroundings in which it finds itself. They have even been known to turn bright red to match floating debris

Fig. 37: A Sea Horse

Seahorse also has a segmented bony armour, anupright posture and a curled prehensile tail. There are about 54 species of seahorses worldwide, and possibly as many more according another publication (Wilson and Orr, 2011). It is often difficult for scientists to identify seahorses because individuals of the same species can vary greatly in appearance. New species continue to be described by marine scientists. "Seahorses have been eaten for more than 2,000 years *(according to ancient records it started way back since 342 BC)* and, as well as being a traditional way of improving sex-drive, are also thought to help respiratory problems and keep one feeling and looking youthful" (Blogspot, 2014). One legendary fan of them (they're fishes, you know!) was Emperor Tangminghuang, one of the most popular emperors of China. He ruled from 712 to 756 and drank sea horse-infused liquor in his later years. This was hundreds of years ago, of course, but the fish remains a best-selling tonic (Asia Obscura 2011).

9.1 INDIA AND SEAHORSE TRADE

Seahorses (Hippocampus spp.) are a major commodity fished from the shallow coastal seas of the south coast of India where there is an abundance of sea grasses, sponges and corals. They are in great demand for export as traditional medicines, curios and aquarium fish. Organised fishing and trade of seahorses exists in India along the Palk Bay and Gulf of Mannar coasts. At the Palk Bay coast, seahorses are targeted by divers along with sea cucumbers (Holothuria spp.) and gastropods (e.g. Murex spp., Xancuspyrum Hornell). In the Gulf of Mannar, most of the seahorses are landed as bycatch of shrimp trawling. Seahorses are also fished from Kerala as a bycatch of trawling, although no organized fishery and trade exists. Five species of seahorses were identified from the Palk Bay coast, whereas only two species were obtained from Kerala. Most seahorses from India are exported to Singapore, Hong Kong, Malaysia and the United Arab Emirates (Salin, Yohannan and Mohanakumaran Nair 2005).

Table 25: Species composition of seahorses landed at various centers of India in 2001
(Ref. Salian *et al.*, 2005)

Landing centers	Species
Thondi *Hippocampusborboniensis Dumeril*	*H. spinosissimus Weber*
	H. kudaBleeker
H. fuscus Ruppell	
H. trimaculatus Leach	
Sakthikulangara, Kochi and	
Calicut	*Hippocampus borboniensis Dumeril*
	H. trimaculatus Leach

Data reproduced with permission of Authors.

According to the reports of National Institute of Oceanography (NIO 2016) 20 million seahorses are traded across the world each year. "India is one of its largest clandestine exporters, shipping 3.6 tonnes or 1.3 million seahorses annually," the NIO told the national government in October. One kilogram (2.2 pounds) or 100 dried seahorses fetch up to 200 dollars in countries such as the United States which have large Chinese populations" (Pratap Chakravarty 2016).

9.2 PECULIARITY OF SEAHORSES BREEDING

All seahorses exhibit male pregnancy (Vincent 1990). When mating female seahorse lays 200-600 eggs. After receiving eggs from females, male seahorses fertilize eggs in specialized abdominal pouches. The eggs develop into embryos therein, under suitable conditions (Linton et.al 1964). Hatching period varies from 3-6 weeks. The young ones are then released to water. Often, they are eaten by predators, mostly fishes. This is a kind biological feature that makes seahorses very appealing to the public and there by emphasizing the importance to conserve them (Scales 2009).

9.3 MEDICAL USES OF SEAHORSE MEAT

Many animals like sea cucumber, octopus, sea horses etc. are eaten by people from ancient times for aphrodisiac properties.

What is an aphrodisiac?

According to Tongkat Ali (2017) an aphrodisiac is defined as *"any substance that can arouse or enhance sexuality and pleasure"*. This substance could be a food, drug, herb, spice, or any kind of animal or plant material. Many scientists and doctors would be surprised to learn that aphrodisiac plants do exist and have been known for millennia. There is such a thing as a love potion formula (more like sex rather than love) but it takes a few days to become most effective, rather than immediately". However, it should be noted that most aphrodisiac have different side effects. Increased blood pressure and heart rate and hypertension are common side effects.

According western medical reports dried seahorses were used 400 years ago to enhance erection, but can treat impotence, asthma, high cholesterol, fiber arteriosclerosis, incontinence, thyroid problems, skin diseases, cardiac diseases, and even broken bones. They are used to treat respiratory disorders, angina, arteriosclerosis, kidney disease, goiter, lymphatic disorders, skin diseases, lethargy, infertility and impotence.

For medicine, seahorses are dried herbs sweet, warm properties is useful and effective in regulating human blood, kidney pitch for men. At the same time, the effects of dry seahorse are also reflected in the crystal cell therapy, knee and back pain, fatigue infertility in women and men. Some people consider seahorses as "Viagra" for men. The hippocampus is also used to treat back pain, pimples. Seahorses are extensively used in several Traditional Chinese Medicines (Chia-Hao Chang 2013) several curative activities of seahorses have been found by researchers. Traditionally seahorses have been used in Chinese medicine to treat erectile disf unction in males (May and Tomoda, 2002). Seahorses are also found to exhibit antiaging, antifatigue properties, suppressneuro-inflammatory responses and collagen release (Ryu *et al.*, 210; Himaya *et al.*, 2012). But most of these applications are popular in Chinese Traditional Medicines and not supported by modern clinical studies. Mostly seahorses are used in drie form.

Today as seahorses are heavily over exploited all around the world. Now, in order to prevent over exploitation and species extinction sea horse is listed an endangered species in several countries and all species of *Hippocampus* were added to Appendix II of the Convention on International Trade in Endangered Species of Wild Fauna and Flora (CITES), effective May 2004.

9.4 HOW TO USE SEA HORSES

Seahorse is sold in markets of Asia mostly as a dried powder, tablet or capsule. Application In the traditional medical use a decoction is prepared by boiling in water. 1 to 3 gms. are used to prepare one decoction. Three doses are prescribed per day. Pregnant women are not recommended to use sea horse (Chen and Chen 2004). In traditional Chinese Medicine seahorse meat is used with several medicinal plants for specific applications.

9.5 REFERENCES

1. Chen JK. And Chen TT. 2004. Chinese Herbology and Pharmacology. City of Industry, C. A: Art of Medicine. Press. P 900.

2. Chia-HaoChang, Nian-Hong Jang-Liaw, Yeong-Shin Lin, Yi-Chiao Fang and Kwang-Tsao Shao.2013. Authenticating the use of dried seahorses in the traditional Chinese medicine market in Taiwanusing molecular forensic. Journal of Food and Drug Analysis 21, 310 e316.

3. Duvernoy, HM. 2005. "Introduction". The Human Hippocampus (3rd ed.). Berlin: Springer-Verlag. p. 1. ISBN 3-540-23191-9. Oxford English Dictionary. Oxford, UK: Oxford University Press. 2007. ISBN 0199206872.

4. Himaya SWA, Ryu B and Qian Z J, *et al.* 2012. Paeonol from *Hippocampus kuda*Bleeler suppressed the neuro-inflammatory responses in vitro via NF-kB and MAPK signaling pathways. Toxicol In Vitro 26: 878e87.

5. Http://aphrodisiaclist.blogspot.in/2014/11/seahorse-aphrodisiac.html.

6. K. R. Salin T. M. Yohanan and C. Mohanakumaran Nair. 2005 Fisheries and trade of seahorses, *Hippocampus spp.,* in southern India. Fisheries Management and Ecology, 12, 269–273.

7. Linton JR, Soloff BL. 1964. The physiology of the brood pouch of the male seahorse Hippocampus erectus. Bull Mar Sci Gulf Caribb ;14: 45e 61.

8. Lourie SA, Stanley HF, Vincent ACJ, *et al.*,1999. Seahorse.

9. May B and Tomoda T. 2002. Seahorses in the Ben cao gang mu and contemporary Chinese medicine. J Aust Chin Med Educ Res Counc; 7:1e 12.

10. Pratap Chakravarty 2016. Yahoo News 1 Dec 06.

11. Ryu B, Qian ZJ and Kim SK. SHP-1(2010). A novel peptide isolated from seahorse inhibits collagen release through the suppression of collagenases 1 and 3, nitric oxide products regulated by NF-kB/p38 kinase. Peptides 31:79e81.

12. Scales H. 2009. Poseidon's steed: the story of seahorses, from myth to reality. New York, USA: Penguin Group.

13. Tongkat Ali.2016.net/aphrodisiacs.htm.

14. Vincent AC. 1990. J. Reproductive ecology of seahorses. Cambridge, UK: University of Cambridge.

15. Why Chinese Pharmacies Sell Dried Sea Horses. 2011, AsiaObscura .com/ 2011/04/strange-tcm—sea-horses-the-oceanic-aphrodisiac.html.

16. Wilson AB, Orr JW. 2011. The evolutionary origins of Syngnathidae: pipefishes and seahorses. J Fish Biol. 78: 1603e23.

CHAPTER **10**

OCTOPUS

Octopuses are caught from the seas by man more than 2000 years ago (Roper *et al.*1984). They are popularly called {"Devil Fishes" and belong to benthic animals that live in coastal seas upto 1000 meters. Around 200 species are identified to exist in the oceans around the world (Worms,1983.). Of these 60 species are reported to occur in the Indian Ocean (Roper *et al.* 1984). The octopus is a wonderful animal living in a universe of its own entirely different from other marine organisms. The arms of octopus are capable of remarkable degree of independent decision making and has the biggest brain compared to any other non-vertebrate animal living in the oceans. The octopus has eight arms and they have got sneakers arranged in rows down the arm.

Fig 38: Octopus alive in natural habitat
Photo courtesy Dr. K. K. IdreesBabu

Octopus dofleini (genus Enteroctopus) are the octopus species which are found in rocky areas, kelp forests and caves in the Pacific Ocean. It could be identified reddish brown or dark red skin. They have eight arms with each arm having 280 suckers. The center of the arms consists of beak and radula which is a toothed tongue. The body is compressible and 9.6 m in radius. It is cool-blooded or poikilothermic having three hearts and blue blood. Mantle is spherical that consists of major organs of this animal. Octopus is able to change the skin's color by adapting the environment due to the presence of tiny pigments called chromatophores. It prefers cold and oxygenated water. They hunt shrimp, scallops, crabs, clams, moon snails, abalones, small octopus, flatfish, rockfish and sculpins for the food. It is the prey for seals, dogfish sharks, sea otters, man, lingcod, sea lions, seals, sea otters, larger octopuses and fish. It has lifespan of 4 to 5 years. Other common names of Octopus dofleini are North Pacific giant octopus, Giant Pacific octopus, Giant octopus, Poulpe géant and Pulpo gigante. Octopusis alsousually found in the Pacific Ocean. It was recorded in Alaskan Aleutian Islands and Baja California. It is extended in Northeast of Japan. The name octopus was kept by Ancient Greeks which means eight feet. It was illustrated on Minoan and Cretan coins and painted on jars in Mycenaenera. Around 300 million years ago, the ancestors of octopus were on Carboniferous seas. The oldest fossil of octopus was found in Chicago of Field Museum. It is distributed to coastal North Pacific in Washington, Oregon, California, Alaska, British Columbia, Northern Japan, Russia and Korea.

According to the report of FAO of UN, 2014 was a bumper year for production of octopus (Global Seafood Marketing, 2015). Total octopus production rose to about 370 000 tonnes, which is the highest level since 2009. The main producer was China, which accounted for over 120 000 tonnes, followed by Japan (ca. 35 000 tonnes) and Mexico (ca. 34 000 tonnes). Interestingly, China did not become a major producer until 2003. Indeed, in 2002 China reported landings of only 741 tonnes. (Globe Fish,01/05/2015). Global octopus landings appear to be on an increasing trend right now. In 2015, landings increased by 6.7 percent compared with 2014. Increases were registered in the major suppliers (Morocco, Mauritania and Mexico), while landings in Spain, Portugal and the Republic of Korea declined (FAO.13/01/2017).

10.1 SEX OF OCTOPUS

Male has got one arm quite different from others. This is the third arm. At the tip of the arm called '*hectocotylus*" is the lingual which serves as a male reproductive organ.A few male octopus lay around one hundred thousand eggs in a season. Once hatched they keep them protected by holding in the lairs for few weeks. Then they are released to in the ocean. Mortality is very heavy as they will be eaten by other fish.

10.2 CANNIBALISM IN OCTOPUS

Unlike many other species male octopuses eat young ones in seasons of non-availability of food in the oceans. This is one reason for their species extinction.

10.3 EYES OF THE OCTOPUS

The octopus eyes are quite different from other marine animals. The eyes of octopus have horizontal pupils. During movements the eyes remain at the same orientation regardless of its position. This facilitates the animal to sight food in all direction and escape from enemies.

10.4 OCTOPUS VENOM

All octopus species contain venom but in different quantities. Only few species have sufficient quantity of venom to kill a man. The blue ringed octopus is one such species that has sufficient venom to kill an enemy. The venom is released to habitat water when the animal is threatened by predators. This is chemically made up of melanin. The colour of the ink varies from red, brown and black. The ink clours the water and frighten the predators and in the meantime the octopus escapes. The dye affects the eye sight of the enemy creating blurred vision and has also a peculiar unpleasant smell.

10.5 OCTOPUS USE IN TRADITIONAL CHINESE MEDICINE

Most of the octopus sold globally is for Traditional Chinese Medicine (TCM) which is now recognized by World Health Organization as a valid form of health care as it is used by one-fourth of the human population. The Traditional Chinese Medicine uses approximately 150 million seahorses per annum. The main use is to prepare aphrodisiac formulations. Dried seahorses are sold in retail markets for prices US$ 600-3000 per kilogram with larger, paler and smoother animals commanding highest prices. Infact, in terms of value based on weight seahorses retail more than the price of silver and almost that of gold in Asia (UNEP, 2004).

The curio and aquarium trade

Because of its colour and shape dried seahorses are used for this purpose. According to International seahorse trust (2006) more that millions of seahorses are used for this purpose. More than one million of seahorses are also used by aquarium trade. (International Seahorse Trust 2007). This indicates the level of over exploitation of seahorses and stressing the need for conservation of this invaluable marine species.

10.6 OCTOPUS FISHERY OF INDIA

Among the cephalopod resources octopuses are least exploited in India although they occur in appreciable quantities around the Indian coastal waters (Silas 1985., Sundaram 2010). There are no exclusive gears to catch octopuses. They usually landed as by catches from shrimp trawlers. According to Nair *et. a/* (1992), the seasons recognized for the cephalopod fishery are the pre-monsoon (February to May), the monsoon (June-August) and thepost monsoon (September-January). The monthly abundance suggests that Octopusfishery is very high during pre - monsoon seasons in Maharashtra i.e., during February-April.

10.7 NUTRITIONAL VALUE

100 grams of cooked octopus contain about 60% moisture,25 to 28 g of protein, and 3g fat. lipid It also a rich source of Vitamin B12, Selenium, Iron, Copper, Magnesium, sodium, and potassium. Octopus is naturally low in fat, but it is high in cholesterol, which can be harmful if you consume too much.

Table 26: Proximate composition of Octopus dofleini

Composition	As % cooked meat in gms.
Moisture	60.50
Lipids	2.08
Protein	29.82
Ash	3.20
Carbohydrate	4.50

Octopus meat is also a good amino acid like leucine, Isoleucine, tryptophan, threonine, lysine, valine, histidine, and vitamins; Vitamin A, Niacin, Vitamin B5, Vitamin B6, Vitamin B12, Vitamin A and Choline. Octopus meat is also a good source for amino acid like leucine, Isoleucine, tryptophan, threonine, lysine, valine, histidine, and vitamins; Vitamin A, Niacin, Vitamin B5, Vitamin B6, Vitamin B12, Vitamin A and Choline.

10.8 Health Benefits

- Treats cancer
- Formation of hemoglobin
- Supports growth
- Hair benefits
- Kidney ailments
- Brain health
- Source of energy
- Reduce stress
- Lowers migraine
- Balance pressure

But none of these benefits are supported by clinical studies.

10.9 OCTOPUS ANATOMY INSPIRED DESIGN OF ROBOTIC ARM

In all animal's neurons are located in the brain. But in octopus the peculiarity is that the nervous system and the density of neurons are mainly located in the tentacles, which taken together, exceeds the total number of neurons in the brain itself. Three-fifths of an octopus's neurons are in its limbs instead of its brain. Consistent with this fact, a recent study showed that a detached octopus arm could be made to flail realistically when stimulated with short electrical pulses... Accordingly, in a detached vertebrate limb it is simply not possible to produce the suite of coordinated movements that is characteristic of complex vertebrate locomotion. In contrast, what is striking about the octopus is the sophistication of the semi-autonomous neural networks in its tentacles and their local motor programs. These observations bear on the assessment of consciousness in the sense that they may alter the notions of embodiment and bodily representation as they have been set forth by cognitive scientists and philosophers. In any case, it is not likely that the question "what it like is to be an octopus tentacle?" will ever be posed by any rational philosopher (Elderman, *et al.* 2005) The octopus arm shows peculiar features, such as the ability to bend in all directions, to produce fast elongations, and to vary its stiffness. The octopus is a boneless animal and it's amazing dexterity is due to its muscular structure where longitudinal (axial), transverse (radial) and oblique muscles seamlessly interact while preserving hydrostaticity i.e. volume conservation (Tao Li *et al*.2012). The octopus achieves these unique motor skills, thanks to its peculiar muscular structure, named muscular hydrostat (. Laschi et.al 2009). Different muscles arranged on orthogonal planes generate an antagonistic action on each other in the muscular hydrostat, which does not change its volume during muscle contractions, and allow bending and elongation of the arm and stiffness variation. By drawing inspiration from natural skills of octopus, and by analyzing the geometry and mechanics of the muscular structure of its arm Laschi *et al.* (2009) designed a robot arm consisting of an artificial muscular hydrostat structure, which is completely soft and compliant.

10.10 REFERENCES

1. C Laschi, B Mazzolai, V Mattoli, M Cianchetti and P Dario, Design of a biomimetic robotic octopus arm, Bioinspiration & Biomimetics, Volume 4, Number 13 Published 4 March 2009 • 2009 IOP Publishing Ltd.

2. C Laschi, B Mazzolai, V Mattoli, M Cianchetti and P Dario, Design of a biomimetic robotic octopus arm, Bioinspiration & Biomimetics, Volume 4, Number 13 Published 4 March 2009 • 2009 IOP Publishing Ltd.

3. David B. Edelman in a 2005 issue of Consciousness and Cognition as cited in Brendan Kiley - Sexy Beast. Anyway, what's it like to be an octopus tentacle?

4. David B. Edelman, Bernard J.Baars. and Anil K.Seth, 2005 . Identifying hallmarks of consciousness in non-mammalian species. Consciousness and Cognition Volume 14, Issue 1, March 2005, Pages 169-187.

5. Https://www.healthbenefitstimes.com/octopus/

6. Nair K.P. Meiappan M.M., Rao G.S., Mohamed K.S., Vidyasagar K., Sundaram K.S. and Lipton .A.P. 1992. Present status of exploitation of fish and shellfish resources: Squid and cuttlefish. Bull. Cent. Ma, Fish. Res. Inst 45: 226-241.

7. Roper, C.F.E., M.J. Sweeney and C.E. Nauen. 1984. FAO Species catalogue. Vol.3. Cephalopods of the world. An annotated and illustrated catalogue of species of interest to fisheries. FAG Fish.Synap. (125) 3: 277 p.

8. Silas. E.G., Sarvesan R., and Rao K.S .1985. Octopod resources. In: Silas, E.G.(Ed.). Cephalopod bionomics, fisheries andresources of the Exclusive Economic Zoneof India. Bull. Cent. Mar. Fish. Res. Inst,37: 137-139.

9. Sujith Sundaram.2010. Octppus Fishery of Indian NW (Maharashtra Coast), Fishing Chimes. Vol. 30, N.8. November, pp. 43-45.

10. Tao Li, Kohei Nakajima, Marcello Calisti, Cecilia Laschi, Rolf Pfeifer, "Octopus-inspired sensorimotor control of a multi-arm soft robot", Mechatronics and Automation (ICMA) 2012 International Conference, pp. 948-955, 2012.

11. Tao Li, Kohei Nakajima, Marcello Calisti, Cecilia Laschi, Rolf Pfeifer, 2012 "Octopus-inspired sensorimotor control of a multi-arm soft robot", Mechatronics and Automation (ICMA) 2012 International Conference, pp. 948-955, 2012.

12. The sea horse dilemma. http://www.saveourseahorses.org.

13. Worms. J. 1883. World fisheries for cephalopods: A synoptic overviewinCaddy, J.E. (Ed.). Advances in Assessment of world cephalopod resources. FAO Fish.Tech. Pap., (231), 1-29

CHAPTER **11**

CONE SNAILS

C one snails, cone shells, or "cones", are common names for a taxonomic family of sea snails known scientifically as the Conidae, marine gastropod mollusks in the superfamily Conoidea. The family Conidae is currently dominated by one genus, Conus, which has about 500 living species. The shells of most of the sea snails in this family are shaped more or less like the geometric shape of a cone, Hence, got the popular and scientific name cone snail. Cone snails are medium-sized to large, sophisticated predatory animals.

Fig, 39: Sea Snails

Scientific classification

- Kingdom   :   Animalia
- Phylum   :   Mollusca
- Class   :   Gastropoda

- (unranked)   :   clade Caenogastropoda
- Clade Neogastropoda
- Superfamily   :   Conoidea
- Family   :   Conidae (Linnaeus, 1758)

11.1 MEDICAL APPLICATIONS OF SNAIL VENOM

Cone snail contains a deadly poison. The cone snail uses this poison, a toxin, filled tooth to harpoon its prey, injecting chemicals that can paralyze, stun or kill an unfortunate fish. Cone snails venom contains toxins which are called conotoxins and now is a medical marvel. These toxins of cone snail are composed of chains of amino acids. In 2003 psychiatrist and environmentalist Eric Chivian of Harvard University described these sea creatures as having "the largest and most clinically important pharmacopoeia of any genus in nature." Scientists believe conotoxins could help treat epilepsy, depression and other disorders by interacting with the nervous system.

Every cone snail species has easily 1,000 peptides of medical interest. Use of these toxins offer immense possibilities for medical research. Cone snail venom contains neurotoxins that can target specific locations in the brain and spinal cord. Some species of cone snail possess a compound that can act on the same receptors as nicotine. These receptors, located on the surface of neurons. Neuroscientist J. Michael McIntosh of the University of Utah has found that selectively blocking some of these receptors with a cone snail compound can decrease the use of addictive drugs (so far, just in laboratory animals). Blocking a different subset of those receptors can trigger more consumption of a drug instead. Other compounds have been found to interact with receptors that influence feelings of pain or the growth of tumors (Scientific American September 2017).

The venom from marine cone snails, used to immobilize prey, contains numerous peptides called conotoxins, some of which can act as painkillers in mammals.Neuropathic pain is a form of chronic pain that occurs in conjunction with injury or dysfunction of the nervous system. This can cause a debilitating and difficult to treat medical condition. According to the research findings of Rittenhouse (2014) neuropathic pain is associated with changes in the transmission of signals between neurons, a process that depends on several types of voltage-gated calcium channels (VGCCs) and cSonotoxins can perhaps can be an inhibitor of this pain.

The venom of some cone snails, such as the Magician cone, *Conus magus*, has been found to be an excellent non-addictive pain reliever 1000 times as powerful as compared to morphine(ANI 2007). Many peptides produced by the cone snails show prospects for being potent pharmaceuticals, such as AVC1, isolated from the Australian species, the Queen Victoria cone, *Conusvictoriae*. This has proved very effective in treating post-surgical and neuropathic pain, even accelerating recovery from nerve injury. The first painkiller Ziconotide derived from cone snail toxins was approved by the U.S. Food and Drug Administration in December 2004 under the name "Prialt". Other drugs are in clinical and preclinical trials, such as compounds of the toxin that may be used in the treatment of Alzheimer's disease, Parkinson's disease, and epilepsy (Louise Yeoman, 2006).

According to a recent report published by the University of Queensland pain treatment researchers have discovered thousands of new peptide toxins hidden deep within the venom of just one type of Queensland cone snail.

Professor Paul Alewood (2015) from UQ's Institute for Molecular Bioscience, said that his team used biochemical and bioinformatics tools to develop a new method to analyze the structure of the venom toxins, allowing them to delve deeper than ever before. According to him "Cone snail venom is known to contain toxins proven to be valuable drug. This study gives the first-ever snapshot of the toxins that exist in the venom of a single cone snail. Cone snail venoms are a complex cocktail of many chemicals and most of these toxins have been overlooked in the past."

Using their new method that involved accurately measuring and analyzing the structure, activity and composition of the diverse range of proteins within venom, researchers discovered the highest number of peptides (mini-proteins) produced in a single cone snail.

"We also discovered six original 'frameworks' — 3D-shaped molecules suitable as drug leads — which we expect will support drug development in the near future," Professor Alewood said.

There are 25 known frameworks discovered over the past 25 years, many of which have already led to a drug or drug lead for several diseases.

"We expect these newly discovered frameworks will also lead to new medications, which can be used to treat pain, cancer and a range of other diseases."

Researchers at UOW and the Illawarra Health and Medical Research Institute (IHMRI) Australia have discovered a new class of molecules originally derived from sea snails that are showing promising results against multi-drug resistant cancers. The molecules, called *N*-alkylisatins, killed 100 per cent of drug resistant cancer cells in the lab in just 48 hours. In comparison, a chemotherapy drug commonly used to treat breast cancer killed only 10 per cent of cells in the same time period (Kara Perrow. 2016). According to this team of researchers *N*-alkylisatins, which proved particularly potent against colorectal, prostate and breast cancers, work by targeting the skeleton of the cell, which is critical for a cell to continue dividing.

The pain-killing compound, called RgIA, is a peptide that naturally occurs in the venom of *Conusregius*, a small sea snail species with a cone-shaped shell, which is common in the Caribbean Sea. According to BaldomeroOlivera, Biologist, University of Utah "Nature has evolved molecules that are extremely sophisticated and can have unexpected applications and they are interested in using venoms to understand different pathways in the nervous system. According to the report (Peter Dockrill, 23 Feb 2017) more than 90 Americans die every day from an opioid overdose, a new treatment formulated from the venom compound could provide an alternative to overused opioid medications – a crisis that's been described as the worst drug epidemic in American history.

However more research has to be done to develop the use of conotoxins therapeutic applications by clinical trials. But research in this field has immense potential in future.

11.2 REFERENCES

1. ANI. 2007. "Sea snail venom paves way for potent new painkiller". Compassionate health care network. Retrieved 2008-11-1

2. Elise Pitt. 2016. Chemical derived from sea snails proving a potent cancer killer:New class of molecules proves promising in the fight against multi-drug resistant cancers. University of Wollongong, Australia.http://media.uow.edu.au/releases/UOW213101.html

3. G. Berecki, J. R. McArthur, H. Cuny, R. J. Clark, D. J. Adams. 2014. Differential Cav2.1 and Cav2.3 channel inhibition by baclofen and -conotoxin Vc1.1 via GABAB receptor activation. The Journal of General Physiology,; 143 (4): 465 DOI: 10.1085/jgp.201311104.

4. Louise Yeoman (2006-03-28). "Venomous snails aid medical science". BBC. Retrieved 2008-11-19.

5. Peter Dockrill. 2017. Sea Snail Venom Could Provide a New, Long-Lasting Alternative to Opioid Painkillers.https://www.sciencealert.com/sea-snail-venom-could-provide-a-new-long-lasting-alternative-to-opioid-painkillers.

6. Piper, Ross . 2007, Extraordinary Animals: An Encyclopedia of Curious and Unusual Animals, Greenwood Press.

7. Rittenhouse. A. R. 2014 Novel coupling is painless. The Journal of General Physiology, 143 (4): 443 DOI: 10.1085/jgp.201411190.

8. Vincent Lavergne, IvonHarliwong, Alun Jones, David Miller, Ryan J. Taft, and Paul F. Alewood. Optimized deep-targeted proteotranscriptomic profiling reveals unexplored Conus toxin diversity and novel cysteine frameworks. PNAS, July 2015 DOI: 10.1073/pnas.1501334112.

CHAPTER **12**

SEA URCHINS

Echinoids or sea urchins (Echinodermata: Echinoidea) constitute a group of exclusively marine invertebrates inhabiting the intertidal down to the deep-sea trenches around the world. Sea urchin fisheries have expanded so greatly in recent years that the natural population of sea urchins in Japan, France, Chile, the northeastern United States, the Canada. World landings of sea urchin, having peaks at 120,000 mt in 1995, are now in the state of decline at present landing are about 82,000 mt only (Aminur Rahman, 2016). However, over half this catch comes from the recently expanded Chilean fishery for Loxechinusalbus. In Europe the important commercial species is Paracentrotuslividus. France and Italy are the important markets. But this species is over exploited and experiencing poor landings in recent years.

Fig. 40: Sea Urchin

12.1 UNIQUENESS OF SEA URCHINS

The sea urchins are found to adapt to changes in sea water carbon dioxide. Due to global warming and fossil fuels carbon dioxide levels in the sea are increasing. Carbon dioxide produces acidity as it makes carbonic acid when dissolved in water. Recently the Stanford scientists have discovered that "some purple sea urchins living along the coast of California and Oregon have the surprising ability to rapidly evolve in acidic ocean water – a capacity that may come in handy as climate change increases ocean acidity. This capacity depends on high levels of genetic variation that allow urchins' healthy growth in water with high carbon dioxide levels" (Rob Jordan 2013). According to their study the long-term environmental mosaic has led to the evolution of genetic variations enabling purple sea urchins to regulate their internal pH level in the face of elevated carbon dioxide.

12.2 SEA URCHINS OF INDIA

Sea urchin *Salmacisvirgulata* (L. Agassiz and Desor 1846) is reported to occur in thePondicherry coast in the sea Bay of Bengal (Satheeshkumar 2011).

Some of the commercially important species seen in India coast are given below:

- *Temnopleuratoreumaticus*
- Echinodiscus spp. (Cake-urchins)
- Echinolampas spp. (Heart sea urchin)
- Stomopneustesvariolaris

There are three types of sea-urchins - the true Sea-urchins which are globular, Cake-urchins with almost circular and flattened bodies, and Heart-urchins shaped like a heart. Sea-urchins are coloured differently in purple, green, ochre and blue. Under each type there are several interesting forms along Indian coasts.

Dried sea urchins enjoy a flourishing trade in the east coast of India. Most important species seen in the Tamil Nadu coast is the species *Temnopleurato-reumaticus.*

The gonad of sea urchin, usually referred to as "Sea urchin Roe", is culinary delicacies in many parts of the world. The roe of sea urchins is considered as a prized delicacy in Asian, Mediterranean and Western Hemisphere countries and have long been using as luxury foods in Japan. Japanese demand for sea urchins raised concerns about overfishing, thus making it one of the most valuable sea foods in the world. The population of the Asian Pacific Region has been using it for long time as a remedy for improving general living tone and treatment for a number of diseases. Sea urchins are eaten by people in the Mediterranean, China and Japan for their intense tasting interiors (actually their sex organs) which are eaten raw.

Gonads of Sea urchin are also rich in valuable bioactive compounds, such as polyunsaturated fatty acids (PUFAs) and β-carotene. PUFAs, especially eicosapentaenoic acid (EPA, C20:5) (n-3)) and docosahexaenoic acid (DHA C22:6 (n-3)), have significant preventive effects on arrhythmia, cardiovascular diseases and cancer. β-Carotene and some xanthophylls have strong pro-vitamin A activity and can be used to prevent tumor development and light sensitivity. Sea urchin fisheries have expanded so greatly in recent years that the natural population of sea urchins in Japan, France, Chile, the northeastern United States, the Canadian Maritime Provinces, and the west coast of North America from California to British Colombia have been overfished to meet the great demand. Not surprisingly, the decrease in supply and the continued strong demand have led to a great increase in interest in aquaculture of sea urchins. Most, if not all, sea urchin fisheries have followed the same pattern of rapid expansion to an unsustainable peak, followed by an equally rapid decline. World landings of sea urchin, having peaks at 120,000 mt in 1995, are now in the state of about 82,000 mt. However, over half this catch comes from the recently expanded Chilean fishery for Loxechinusalbus. In Europe, the sea urchin stocks (Paracentrotuslividus) of first France and then Ireland were overfished in the 1980s to supply the French markets. However, these decreasing patterns clearly reflect the overexploitation of most fishery grounds and highlight the need for aquaculture development, fishery management and conservation strategies. While the wild stocks decline, high market demand for food, nutraceuticals and pharmaceuticals, increases the price of the product and thus, culturing is most likely to become commercially viable. As this review shows, there have been dramatic progresses in the culture techniques of sea urchins in the last 15–20 years; we can conclude that currently the major obstacles to successful cultivation are indeed managerial, cultural, conservational and financial rather than biological. Therefore, the fate of the sea urchin fishery is closely connected to that of the fisheries, whose fortune will ultimately depend upon the stock enhancement, culture improvement, quality roe production and market forces that will ultimately shape this growing industry in a sustainable manner.

Table 29: Commercially important species of sea urchin in the world

Local name	Scientific name	Place of habitat
Sea urchin	*Tripneustesgratilla*	Hawaii, and the Red Sea, Mozambique, Red Sea, westward to Hawaii and Clarion Island
Kina	*Evechinuschloroticus*	New Zealand
Purple sea urchin	*Heliocidariserythrogramma*	Australia
Chilean Red Sea Urchin,(erizo)	*Loxechinusalbus*	Pacific coast of South America
Sea urchin	*Lytechinusvariegatus*	Atlantic, from Florida, through the Caribbean to Brazil
Common sea urchin	*Paracentrotuslividus*	Mediterranean and the north east Atlantic.
Green sea urchin	*Strongylocentrotusfranciscanus*	West Coast of North America from Baha California to the Aleutian Archipelago and the coast of Siberia and North A
Green sea urchin	*Strongylocentrotusintermedius*	Asian and Siberian coast of the Pacific, Japan.
Japanese sea urchin	*Strongylocentrotusnudus*	China northwards to PrimorskyiKray, Russia and in Japan
Purple sea urchin	*Strongylocentrotuspurpuratus*	Pacific coast of North America
Purple crowned urchin	*Centrostephanusrodgersii*	Australia and New Zealand
White sea urchin	*Salmacissphaeroides)*	Indo-West Pacific from China to Solomon Islands and Australia, and Singapore
India sea urchin	*Temnopleuratoreumaticus*	West coast of India 9 Bay of Bengal
Indian sea urchin	*Stomopneustesvarolaris*	Bay of Bengal, India.

Proximate composition varies from species to species, seasons, spawning, sea water temperature and seasons. Availability of the feed also is a major factor. At best one can give only an approximate figure. The gonads of sea urchin are also a rich source of essential amino acids and polyunsaturated fatty acids (Ayyagari, Archanaand, Ramesh Babu 2016).

Table 28: Proximate composition of Indian sea urchin (*Stomopneustes variolaris*) Dry weight basis (g/100g dried meat)

Composition	Dry weight basis
Moisture	77.5-76
Ash	3.7-4
Protein	12-13
Lipids	4-5
Carbohydrates	1.5-2

Table 29: Nutrients in sea urchin meat (wet body weight)

Composition	Dry weight basis
Calories	172 kcal/100 g
Protein	13.27/100g
Polyunsaturated fatty acids	1.75/100g
Omega 3 fatty acids	1.07/100g
Eicosapentaenoic acid (EPA)	0.79/100g
Docosahexaenoic acid (DHA).	0.04/100g
Zinc	17.00 ppm

12.3 Products from sea urchins

There are a variety of sea urchin recipes viz., Risotto with Sea Urchin-Dill, and Smoked Caviar-Sea Urchin Mousse with Ginger Vinaigrette, Sea Urchin Bruschetta and Sea Urchin Linguine available in the western world. In India not all the coastal community eat the gonads of sea urchin, but fishers of a fishing village along Gulf of Mannar, have the habit of consuming the sea urchin roe for centuries (Saravanan *et al.*, 2016).

Sea urchins are eaten by the coastal fishermen of the East coast of India. Heat processed (grilled) and cooked sea urchins are the two processing methods practiced by the for centuries.

Grilled seaurchin roe: The sea urchin (Stomopneustesvaroluris) is collected and cleaned well. See that its spine is kept intact. Heat on fire, normally on fire wood, for 30-40 minutes. Cooking is over. The roe is ready for consumption.

Cooked-dried sea urchin: Clean the sea urchin well in potable water. Take sea urchin gonads in a cooking vessel with sufficient water. Add some Moringa leaves (250g/750g gonads). Stir constantly. Heat on fire for 30 minutes. Dry the cooked meat in the sun. Some tomes the product is smoked also.

Table 30: Nutrient Profile of dried sea urchin gonads (dried weight basis) Ref. Saravanan *et al.* (2016).

Composition	Dry weight basis
Protein	40%
Fat	29%
Crude Fiber	0.3%
Nitrogen free extract	22%
Total ash	8%
Calories	5000 Kcal./Kg

The urchins are exploited commercially for extraction of gonads which are used as food and nutraceutical. Following species are commercially important for production of gonads;

Genera: *Centrostephanus, Diadema, Arbacia, Echinus, Loxechinus, Paracentrotus, Psammechinus, Anthocidaris, Colobocentrotus, Echinometra, Evechinus, Heliocidaris, Hemicentrotus, Strongylocentrotus, Lytechinus, Pseudoboletia, Pseudocentrotus, Toxopneustes,* and *Tripneustes. (Sloan,* N.A. 1985; Saito, 1992; Keesing and Hall, 1998; Lawrence, 2007).

Sea urchin is better known in Japan as "uni". Sea urchin is a tasty and highly nutritionally rich food with proteins and minerals. It is also delicious. Red sea urchins are the largest sea urchin. It is found only along the rocky sub-tidal Pacific shores of North America. They are harvested for their soft tissues known as "uni", a highly valued seafood delicacy. Uni is used to prepare the best quality sashimi a Japanese delicacy. One hundred gram of sea urchin gives nearly 120 calories. It is also a rich source of omega-3 fatty acids. 100-gram sea urchin meat gives 13.3 gm proteins. Hence it is rich source of proteins (Kaneniwa and Takagi 1986., Ichihiro, 1986).

12.4 MEDICAL APPLICATIONS OF SEA URCHINS

Among the health benefits of eating sea urchin include improved virility among other benefits. Sea urchin is rich in proteins. It is thus a good substitute for protein rich foods such as meat, fish, beans and legumes. And a good source for dietary fiber.

Sea urchins is a good source supply of vitamin C and A. Vitamin C is essential to your body for combating deficiencies such as scurvy and Vitamin A to boost Immune System

In many South-East Sian countries like Japan, Korea and China sea urchin roes are a popular aphrodisiac to improve libido and virility (Yur'eva *et al.* 2003). The sea urchin roes (spawn) improves blood circulation of the reproductive organs. Hence, it will improve on your sexual arousal. This could also be due to high zinc content (17 ppm/100g meat) in the meat. But there are no clinical studies supporting these claims.

12.5 REFERENCES

1. Aminur Rahman M, 2016. World Sea Urchin Fisheries: Their Current Status, Culture Practices, Management Strategies and Biomedical Applications. Conference on Aquaculture & Fisheries, Harvest Future Sustainable Aquaculture July 11-13, 2016 Kuala Lumpur, Malaysia

2. Ayyagari Archana K. Ramesh Babu 2016, Nutrient composition and antioxidant activity of gonads of sea urchin *Stomopneustes variolaris,* Food Chemistry Volume 197, Part A, pp.597-602

3. Ichihiro, K. 1986. Breeding, processing and sale, Hokkai Suisan Shinbunsha, Sappro, Japan.

4. Kaneniwa, M. and Takagi, T. 1986. Fatty acids in the lipid of food products from sea urchin. Bulletin of the Japanese Society of Scientific Fisheries, 52(9): 1681–1685.

5. Keesing, J.K., Hall, K.C. 1998. Review of the status of world sea urchin fisheries points to opportunities for aquaculture. Journal of Shellfish Research, 17(5):1597–1604.

6. Lawrence, J.M. 2007. Edible Sea Urchins: Biology and Ecology. Elsevier, Boston, 380 p.

7. Rob Jordan 2013.Stanford seeks sea urchin's secret to surviving ocean acidification. Stanford News services, April 8 2013.

8. Saito, K. 1992. Sea urchin fishery of Japan. In: Anonymous, The management and enhancement of sea urchins and other kelp bed resources: a Pacific rim perspective. California Sea Grant College, La Jolla. Rep. No. T-CSGCP-028.

9. Saravanan R, K.K. Joshi, I. Syed Sadiq, A.K. Abdul Nazar 1 and S. Chandrasekar. 2016, Traditional Knowledge on the Edibility of Sea urchin roe among the Fisher Folk Community of the Gulf of Mannar region with a note on their Cuisine. 1si international Agro-Biodiversity Congress. Science, Technology, Policy and Partnership. November 6-9, New Delhi, 2016, India

10. Satheeshkumar P. 2011 First Record of Regular Sea Urchin Salmacisvirgulata (L. Agassiz and Desor 1846) from the Pondicherry Coast, India. World Journal of Fish and Marine Sciences 3 (2): 126-128, 2011 ISSN 2078-4589.

11. Sloan, N.A. 1985. Echinoderm fisheries of the world: a review. In: Keegan, B.F. and O' Connor, B.D.S. (Eds.), Echinoderms. A.A. Bolkema, Rotterdam, pp. 109–124.

12. Yur'eva, M.I., Lisakovskaya, O.V., Akulin, V.N., Kropotov, A.V. 2003. Gonads of sea urchins as the source of medication stimulating sexual behavior. Russian Journal of Marine Biology, 29: 189–193. 9.

CHAPTER **13**

SEAL PENIS

Seal most commonly refers to as Pinniped belongs to a diverse group of semi-aquatic marine mammals. They are commonly called seals. The word Pinnipeds (/Èpjnjìp[dz/) combines from two Latin words pinna "fin" and pes, pedis "foot"(Elias, 2007). The seals are a widely distributed and diverse clade of carnivorous, fin-footed, semiaquaticmarine mammals (Berta and Churchill 2012). They comprise the extant families Odobenidae (who's only living member is the walrus), Otariidae (the eared seals: sea lions and fur seals), and Phocidae (the earless seals, or true seals). There are 33 extant species of pinnipeds, and more than 50 extinct species have been described from fossils. Historically seals arethought to have descended from two ancestral lines but molecular evidencesupports them as a monophyletic lineage (descended from one ancestral line). Pinnipeds belong to the order Carnivora and their closest living relatives are bears and musteloids (weasels, raccoons, skunks, and red pandas), having diverged about 50 million years ago. (Wikipedia)

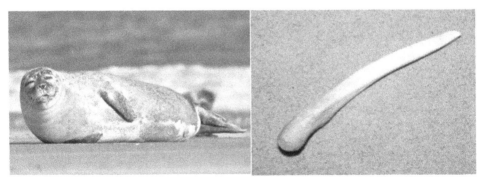

Fig. 41: Seal & Seal Penis bone (Pinnipedia spp.)

13.1 HISTORY AND EXPLOITATION

In ancient Greece, monk seals were placed under the protection of Poseidon and Apollo because they showed a great love for sea and sun. One of the first coins, minted around 500 BC, depicted the head of a monk seal, and the creatures were immortalized in the writings of Homer, Plutarch and Aristotle. To fishermen and seafarers, catching sight of the animals frolicking in the waves or loafing on the beaches was considered to be an omen of good fortune

Some of the important Seal Species

- Leopard Seal
- Harp Seal
- Harbor Seal
- Gray Seal
- Northern Elephant Seal
- Southern Elephant Seal
- Weddell Seal
- Hawaiian Monk Seal
- Mediterranean Monk Seal

13.2 SEAL HUNTING

Seal hunting, or sealing, is the personal or commercial hunting of seals. In the past seal hunting was practiced in eight countries and one region of Denmark: Canada, the United States, Namibia, Iceland, Norway, Russia, Finland, Sweden, and Greenland. Most of the world's seal hunting takes place in Canada and Greenland. Canada's largest market for seals is Norway. Today, commercial sealing is conducted by only five nations: Canada, Greenland, Namibia, Norway, and Russia. The United States, which had been heavily involved in the sealing industry, now maintains a complete ban on the commercial hunting of marine mammals, with the exception of indigenous peoples who are allowed to hunt a small number of seals each year. Norway has fixed a quota annually for seal hunting. There is now a growing global concern on seal hunting as many species are facing stock depletion.

13.2.1 Aphrodisiac properties of seal penis

Dried seal penis is used to prepare a variety of medicinal products to improve male erectile disfunction. But this is a property based largely on a belief and not supported by clinical studies. But several hundred-sex stimulants available in the market and flourishing trade of seal penis tonic is still a mystery. If it does not have some activity the trade would have never survived. Every year thousands of seals are captured and used to extract penis and kidneys. According to report of Department of Fisheries Government of Canada "Asian consumers, particularly athletes, also consume a beverage called Dalishen Oral Liquid that is made from seal penis and testicles, which they believe to be energizing and performance enhancing." They aren't, of course, but honesty is not a hallmark of the fur industry.

According to a recommendation of the Fur Institute of Canada made the proposal of killing 140,000 grey seals over a five-year period in the southern Gulf of St. Lawrence. Grey seals are reported responsible for the catastrophic decline in certain fisheries, especially the northern cod. This is one reason the DFO of Canada has sanctioned hunting of140,000 grey seals over a five-year period in the southern Gulf of St. Lawrence. Whatever be the actual fact with regard to the aphrodisiac properties of seal penis, it enjoys a flourishing trade in many Asian countries. "Asian consumers, particularly athletes, also consume a beverage called Dalishen Oral Liquid that is made from seal penis and testicles, which they believe to be energizing and performance enhancing" (MacKay, Born Free USA, June 2015).

A product called a *Jiftip* in use by men to seal their urethras shut before sex, supposedly in order to protect their penises and stop anything getting out, has become very popular in the market. It is out of seal penis. It is not a one hundred percent safe to avoid pregnancy. But it is still in use. This is used by man to control erectile disfunction.

13.3 REFERENCES

1. Baljinder Singh, VikasGupta, ParveenBansal, Ranjit Singh and Dharmendra Kumar. 2010 Pharmaceutical potential of plant used as aphrodisiacs. International Journal of Pharmaceutical Sciences Review and Research. International Journal of Pharmaceutical Sciences Review and Research, Volume 5, Issue 1, November- December. Pp 104-112.

2. Berta, A.; Churchill, M. 2012. "Pinniped taxonomy: Review of currently recognized species and subspecies, and evidence used for their description". Mammal Review. 42 (3): 207–34.

3. Elias, J. S.2007. Science Terms Made Easy: A Lexicon of Scientific Words and Their Root Language Origins. Greenwood Publishing Group. p. 157.ISBN 978-0-313-33896-0

4. "Seal hunt set to resume off Newfoundland's coast". CTV News. 2006-04-12.

5. https://en.wikipedia.org/wiki/Pinniped.

Chapter **14**

ABALONE

A balones are reef- dwelling marine snails belonging to the Phylum Mollusca, Class *Gastropoda*, Family *Haliotodea* and Genus *Haliotis (Gofas et al., 2014). Abalones are single-shelled* and herbivorous animals widely distributed in the coastal seas. About one hundred species are identified so far occurring in both tropical and subtropical seas around the world. Abalone habitat of temperate water grows to larger size compared to tropical species which are smaller in size. (Estes *et al.*2005). All abalones are benthic rock dwellers, moving relatively short distances throughout their life span. While some species may migrate from deep to shallow depths in search for food, others spend their whole life span on the same home ground. Each species of abalone has a different depth and latitudinal distribution.

Abalone's are often dubbed "white gold" after the pearly underflesh of the snails. They're relished in restaurants in China and all over South eastern countries in Asia. Today abalone trade is a multibillion dollar global export industry. Of the 56-global species South Africa, and *Haliotismidae*, is regarded as among the tastiest.

Some of the commercially important species are given below;

● Black abalone (*Haliotis cracherodii*)

● Green abalone (*Haliotis fulgens*)

● Pink abalone (*Haliotis corrugata*)

● Red abalone (*Haliotis rufescens*)

● White abalone (*Haliotis sorenseni*)

The only identified species of of abalone in India is *Haliotisvaria,* which grows to a maximum length of 8 cm. It is distributed in the gulf of Mannar and

Andaman and Nicobar Islands. (Najmudeen and Victor, 2004). There is no commercial exploitation of abalone so far in India.

14.1 ABALONE FISHERY

In the past abalone was mostly harvested from the seas. But because of its global ever-increasing demand production of abalone was started by farming. In 2014 production of abalone by farming was estimated to be about 103, 464t. China, Korea and Australia are the major producers by farming. But today Australia supplies more than 505 of the abalone from wild stocks of the Global Market. (Cook, 2010). China is the world's biggest consumer of abalone followed by Japan and European countries. *H. discushannaiis* the species that commands highest commercial price in the global market. Japancse consume abalone in fresh form. It is used in the preparation of Sushi a Japanese high-priced delicacy. Abalone is also used for the production of frozen fillets and also as a canned product. Canned abalone is a low-priced product Fresh abalone is graded in three categories in the Japanese markets and they are;

- 9 cm - 9-10 nos. in 1 kg.
- 6.5 cm – 15- 20 nos., in 1kg.

14.2 PROCESSING OF ABALONE

The abalone take seven years to reach maturity and are then harvested, processed, and packaged for export in different forms: dried, canned, and frozen. High quality abalone today fetches a commercial price over US $ 200 per pound. Today it is also sold in live form.Processed abalone can be frozen at -40^0 C as block form or as IQF and marketed.

14.3 MEDICAL APPLICATIONS OF ABALONE SHELL (SHI JUE MING)

The inner shell has keratin with a pearl-like luster and it produces 16 kinds of amino acids after hydrolysis by hydrochloric acid. Main chemical constituents of abalone shells are calcium carbonate and organic matter. It also contains a small amount of magnesium, iron, silicate, phosphate, chloride and a trace amount of iodine, zinc, manganese, chromium, strontium, copper and other trace elements. When calcined the organic matter is completely destroyed, and residue contains mainly calcium oxide.

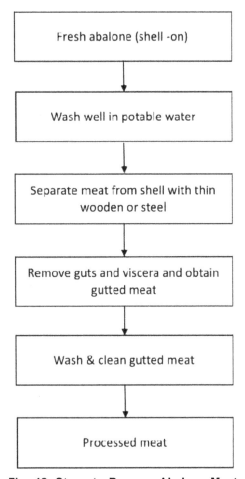

Fig. 42. Steps to Process Abalone Meat

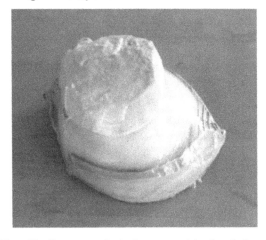

Fig. 43: Processed abalone meat in fresh form

Fig. 44: Abalone shell

14.4 Abalone shell health benefits

The Chinese Materia Medica says that it is salty in flavor and cold in nature and it goes to liver meridian. Essential functions are calming the liver, clearing heat, improving eyesight, and removing nebula. Major abalone shell uses include headache, dizziness, red eyes, nebula, dim vision, glaucoma, and night blindness. Recommended dose is from 10 to 30 grams in decoction. It is often used in the form of pills and powder also. The powdered form is the preferred one as it can immediately have dissolved in the stomach.

In the west, it is mostly used in art industry to make seashell crafts or jewelry, such as abalone shell engagement ring, beads, necklace, buttons, bracelet, cabochon, hair clips, pendant, and so on. But in the east, especially in China, it is a valuable medicine, which is listed as a top-grade herb in the Ming Yi Bie Lu (Miscellaneous Records of Famous Physicians). In the past this herb was scarce due to the scarcity of abalone and overfishing. But today it is no longer rare with the development of aquaculture.

As a matter of fact, it is becoming a part of the diet & medication for coastal people.

The inner surface of the abalone shell emits dazzling multicolored pearl-like luster, which is the secreted abalone nacre and which is supposed to have amazing healing properties according to Chinese herbal medicines. In the Chinese medicine it is used to heal many disorders such as flaming up of the liver-fire and high blood pressure mostly caused by hyperactivity of liver-yang. The corresponding remedy is to calm the liver and suppress yang. Actually, that is where it gets in since it is considered an effective drug of clearing liver and improving vision. That explains well why it is commonly used in the treatment of hypertension-induced dizziness, headaches, and the like. From the modern medical analysis, its main active ingredients are calcium carbonate, keratin, and choline contained in nacre. More importantly, the keratin can be hydrolyzed by stomach acid into 16 kinds of amino acids, which is the equivalent of parasympathomimetic drugs that can reduce visual disturbances and urinary discomfort, and lower blood pressure.

Both abalone shell and cassia seed can treat liver-heat induced red painful swollen eyes and a faint cloudy spot on the cornea by removing liver-fire for improving eyesight. However, these are two different Chinese herbs though both of them are called "Jue Ming". The former is salty in flavor and cold and heavy in nature, which makes an excellent herb of nourishing liver yin, cool liver, and tranquilizing liver. So, it is suitable for both excess and deficit types of eye problems and mostly used in photophobia, dim eyesight, glaucoma, and more. In comparison, the latter is bitter in flavor and cold in nature, which makes it an herb that is better at clearing liver-heat for improving vision. Therefore, it is more commonly used for the red painful swollen eyes due to excess fire in liver channel.

14.4.1 Abalone Visceral Polysaccharide (AVP)

Abalone viscera contains several polysaccharides. They find extensive application healthcare products. But one of the major problems with abalone viscera is that it contains several toxic heavy metals causing health hazards. The most common metals seen in AVP are cadmium, lead, arsenic, chromium, nickel and mercury. Recently a method was developed to remove these metallic residues from AVP by Renjie Xu *et al.* (2016). This new method to extract abalone visceral polysaccharide (AVP) and remove its heavy metals (except for mercury) employs combined use of enzymatic hydrolysis, plate frame filtration, alcohol precipitation, high activated clay absorption with cation exchange resin and anion exchange resin. According to the authors this method gives high quality AVP with low limits of metallic residues and can be safely used by food industry to develop AVP into healthcare products and researchers in the production of abalone visceral polysaccharide.

14.4.2 Some pharmacological uses of haliotis shell

- The extract of *Haliotis diversicolor* (Reeve) has antibacterial effect. The hydrolysate of its inner shell has hepatoprotective effect. This has been confirmed by the resistance experiments of carbon tetrachloride induced acute poisoning in mice;

● Its acidic extract showed significant anticoagulant effect *in vivo* and *in vitro*, which has been indicated by clotting experiments in rabbits.

Abalone shells recipes on herbal remedies

● E Jiao JiZi Huang Tang from Tong Su Shang Han Lun (Popularized Treatise on Cold Damage). It is formulated with Bai Shao (White Peony Root), Sheng Di Huang (RehmanniaGlutinosa), Mu Li (Oyster Shell), etc. to treat heat scorching kidney yin causing contracture of tendon, a slight twitching in hand and foot, dizziness, and the like.

● Ping Gan Qian Yang Tang from Chang Jian Bing Zhong Yi Zhi Liao (Traditional Chinese Medicine Treatment and Research on Common Diseases). It is combined with Xia Ku Cao (Prunella), Huang Qin (Scutellaria), JuHua (Chrysanthemum Morifolium), etc. to cure the sole liver-yang hyperactivity induced dizziness, headache, and irritability.

● Huang Lian Yang Gan Wan from QuanGuoZhong Yao Cheng Yao Chu Fang Ji (The National Prescription Collection of Chinese Patent Medicine). It is matched with Huang Lian (Coptis Root), Long Dan Cao (Gentian Root), Ye Ming Sha (bat dung), etc. to heal red painful swollen eyes due to liver fire flaming.

● Shi Jue Ming San from ZhengZhiZhun Sheng (The Level-line of Patterns and Treatment). It is equipped with Mu Zei (Horsetail Herb), Jing Jie (Schizonepeta), Sang Ye (Mulberry Leaf), BaiJuHua (White chrysanthemum), Gu Jing Cao (Eriocaulon, Pipewort), Cang Zhu (Atractylodes), etc. to treat corneal clouding.

● Shi Jue Ming San from Sheng JiZong Lu (Complete Record of Holy Benevolence). It is joined with QiangHuo (Notopterygium), Cao Jue Ming (Foetid Cassia Seeds), chrysanthemum, and Gan Cao (Licorice Root) to cure dim vision and dizziness due to wind-pathogen attacking head.

● Qian Li Guang Tang from Yan Ke Long Mu Lun (Secret Ophthalmology of Nagarjuna). It is used along with Hai Jin Sha (Lygodium), Licorice Root, and chrysanthemum to heal photophobia.

14.5 COMMERCIAL ABALONE PRODUCTS SOLD IN THE MARKET

- Dried abalone
- Chilled abalone meat
- Abalone packed in Retort Pouches.
- Canned Abalone (normally in brine)
- Vacuum packed Abalone
- Individually Quick Frozen (IQF), shellon.
- Individually Quick Frozen (IQF) meat.
- Live abalone in sea water at low temperatures packed in boxes.

Dried abalone

Abalone is a luxury shellfish that is often served at Chinese weddings and other celebrationsDried abalone has an acquired, expensive taste, like caviar or truffles and is used to impart a distinct flavor to Chinese soups. Dried abalone meat has its own unique characteristics, a sweeter taste and firmer structure than canned and frozen ones. Dried abalone connoisseurs enjoy abalone by braising it whole and eating it like a steak. Dried abalone can be added in soup. Canned or fresh abalone has far less concentrated flavor. It is also eaten after simmering in broth for several hours to soften and served whole or in slices with a savory sauce.

Fig. 45: Dried abalone

Drying method of Abalone (South African Method)

There are many different methods of drying an abalone, but the most extensively applied method is the sun drying method. The drying process is also an art. It is extensively practiced in South Africa. The sun drying process takes up to two months.The abalone is normally shucked, and later on is simmered in preservation chemicals, and hung on racks in a room heated up to 38 degrees Celsius. It is left there to dry for a period of three to four weeks.

Dried Abalone can be kept stored for months or even years. Later it on can be rehydrated and returned to its original form. This method is vital and is usually used in smuggling methods. Primarily, live or frozen abalone carries a pungent and typical smell hence is tricky transport and ship them unnoticed.

Quality of Abalone of different types of Drying methods

Sunxian Zhang and Baodong Zheng (2013) did a detailed study on the quality of abalone dried by sun drying (SD), freeze drying (FD), hot-air drying (HAD), and microwave-vacuum drying (MV). The quality parameters were compared according to the rehydration, color, texture, and content of amino acids. The cohesiveness, viscosity, and chewiness of the products dehydrated by MVD were much better than in the other methods. The color of these products was mostly accepted by the consumers. Meanwhile, the rehydration quality of the MVD products was only secondly to that of the FD samples, and the amino acids in the products were more even distributed than in the others. Consequently, the optimal drying method for the abalones was MVD. It is a well-established fact that freeze drying is by far the best drying technique in terms the quality of dried products compared to all methods. However, the cost is too high, and an average consumer cannot afford to purchase freeze did abalone. Sun dried form is by far the best commercially viable technology.

14.6 REFERENCES

1. Gofas Serge, Tran BastienandBouchetPhillippe "WoRms Taxon Details: Haliotis Linnaeus, 1758". WoRMS (World Register of Marine Species), Revised August 16, 2014.

2. *Gofas, Serge; Tran, Bastien; Bouchet, Phillippe (2014).* "WoRms Taxon Details: Haliotis Linnaeus, 1758". *WoRMS (World Register of Marine Species).* Archived *from the original on 16 August 2014.* Retrieved August 16, 2014.

3. http://food-drink.blurtit.com/1308277/how-to-dry-and-preserve-abalone.

4. http://www.chineseherbshealing.com/abalone-shell.

5. Najmudeen, T. M., and Victor., A. C. C.. Reproductive Biology of the tropical Abalone *Haliotisvaria* from Gulf of Mannar. J. mar. boil. Ass. India, $6 (2): 154-164,July-Dec.,2004.

6. RenjieXu, MeixiangChen,Ting Fang and Jinquan Chen. A New Method for Extraction and Heavy Metals Removal of Abalone Visceral Polysaccharide. *Journal of Food Processing and Preservation.*, First published: 9 August 2016DOI: 10.1111/jfpp.13023

7. Sunxian Zhang and Baodong Zheng Effect of drying methods on quality of abalone. Journal: Food, Agriculture and Environment (JFAE) Online ISSN: 1459-0263Year: 2013, Vol. 11, Issue 3&4, pages 444-447.

CHAPTER **15**

CEPHALOPOD INK

Cephalopoda belong to a class called Mollusca. It is now considered to have lived in the oceans more than 500 million years ago. The Cephalopods are represented by two major extant groups: Nautiloidea (nautilus) and Coleoidea. The coleoids are further classified into Octopodiformes (Vampyromorpha, vampire squids), Octopoda, (octopuses) and Decapodiformes (squids and cuttlefishes). There are more than 700 species cephalopods identified today worldwide and are widely distributed in many different oceanic habitats (Charles 2014).

Aristotle wrote about inking cuttlefish 2500 years ago that "When the Sepia is frightened and in terror, it produces this blackness and muddiness in the water, as it were a shield held in front of the body in fact, A himself has been described by his contemporaries and modern scholars as cephalopodan, with the simile that "he is like the cuttlefish who obscures himself in his own ink when he feels himself about to be grasped" (Schmit.1965).

15.1 INK

All orders of the Coleoidea have ink sacs and produce ink. But the members of the Nautiloidea do not have ink sac. The ink is produced and used by many cephalopod species living in low-light or dark conditions, including the deep sea (Bush and Robison 2007). The ink sac is present at hatching, so even at a small size and young age, cephalopods can produce and release ink (Von Boletzky 1987). It is used as a defense mechanism against enemies. Cephalopod ink is composed of secretions from two glands. The ink sac with its ink gland produces a black ink containing melanin. The ink gland releases its secretion into the ink sac lumen, where it is stored and eventually released via a duct into the hindgut near the anus. The release of ink from the ink sac is controlled by its muscular walls and a pair of sphincters.

A second organ called the funnel organ is a mucus-producing gland. Several researchers had investigated on the structure and role of this organ (Müller 1853, Verrill 1880, Hoyle 1886, Laurie in 1888, Weiss 1888, Brock 1888, Voss 1963 and Hu *et al.* in 2010). The funnel organ, due to its size and location, appears to be the only mucus gland that could secrete this volume of mucus and co-release it with the ink sac's secretion (Charles 2014). Several scientists have suggested that that the combined secretions of the ink sac and funnel organ, produced in different amounts, leads to ink of different forms, ranging from a diffuse cloud to a discrete object with the general appearance of a squid, and, thus, called a pseudomorph, to longer and thinner forms, called ropes (Bush and Robison 2007, Von Boletzky 1987,1997; Young and Mangold 2014)

15.2 COMPOSITION OF SQUID INK

Cephalopod ink is a combination of secretions from ink sac and funnel. Not much detailed studies are reported for the funnel ink. But there are several studies conducted on sack ink. One ink sac of *Sepia* contains about 1 g of melanin (Prota *et al.,* 1981). Melanin constitutes about 15% of the total wet weight of ink (Wang *et al.,* 2014). Proteins make up another 5%–8% of the weight of *Sepia* ink (Prota 2000).

Squid ink is a mixture of proteins, lipids, glycoaminoglycans and several minerals. The main components are protein-polysaccharide complexes and melanin. The ink has been shown to be effective as a bioactive material in several medical applications. Squid-ink containing polysaccharides have shown to inhibit angiogenesis (birth of new blood vessel). This property can suppress the growth of tumors and hence prevent cancer development.

15.2.1 Constituents of squid ink

Several biologically important constituents are seen in the ink. Following are some of them:

- **Enzymes:** The ink contains Tyrosinase, Dopamine re arranging enzymes and Peroxidases as major enzymes
- **Amines:** The principal amine seen in the cephalopod ink is Catecholamines.
- **Peptidoglycans**

- **Amino acids:** The ink is rich in most essential amino acids. However, Taurine is present in high levels about 50% of the total amino acid pool, followed by tyrosin (7%).

- **Toxins:** Generally, cephalopods like squid does not contain toxins. But they can occur from the food they consume.

- **Metals:** Cephalopod like squid and cuttle fish contain high levels of cadmium and copper.

Melanin is the most important pigment seen in cephalopod inks. It is produced from the amino acid tyrosine by a series of biochemical conversions. Several enzymes are involved in the process as well as copper and iron. The principal enzyme involved is tyrosinase normally present in crustaceans and cephalopods (Prota *et al.,* 1981). The biosynthesis of melanin is initiated by the catalytic oxidation of tyrosine to dopabyenymetyrosinase in a reaction that requires dopa as a cofactor. Tyrosine then catalyzes the dehydrogenation of dopa to dopaquinone. The dopaquinone is finally polymerized to melanin.

Melanin is composed of two biopolymers, eumelanin and phenomelanin. Eumelanin is the form seen in cephalopods ink (Palumbo 2003). Most of the studies on cephalopods ink are done in cuttle fish ink (*Sepia officinalis*). Sepia melanin is found to exist as several particles of size varying from 100-200 M (Palumbo, 2003; Clanc and Simon 2001; Matsuura, 2009).

15.3 USE OF INK BY CEPHALOPODS

Ink is used as a defense mechanism against predators. When the ink is jetted out of the body it colours the nearby water body and also brings out changes in body coloration, such as in the blanch-ink-jet maneuver (Hanlon and (Messenger 1996). Ink clouds could serve as smoke-screens behind which inking cephalopods can hide or jet escape. It also affects the the vision of the predators giving time to escape from the invaders. Bush and Robison(2007) observed that deep-sea *Heteroteuthis* squid have ink sacs producing an ink containing mucus and luminescent bacteria from their light organs and that create luminescent clouds, which the squid might use as a visual defense to either conceal themselves or confuse predators According to Bush and Robison [2007] six types of ink are released from deep-sea squid, which they classify as pseudomorphs, pseudomorph series, ink ropes, clouds/smokescreens, diffuse puffs and mantle fills.

15.4 MEDICAL USES OF CEPHALOPODS INK

In recent times cephalopods ink has become a subject of investigation for its clinical applications In Chinese traditional medicine several health benefits to man has been attributed to cephalopod ink (Zhong *et al.,* 2009). Nair *et al.* (2011). has given an extensive review on the medical uses of ink of cephalopods. But much of these uses are not been proved by clinical studies or approved by certifying agencies. Some the important clinical uses given to the ink are given below:

- Anti-microbial properties.
- As anticancer agent
- Antiulcergenic agent
- Used to lower hyper tension
- Hematopoetic effects
- As antioxidant.
- As anti-inflammatory agent.

15.5 USE IN FOOD INDUSTRIES

In recent times cephalopod inks, particularly squid ink is used in food processing as a natural colouring agent. Commercially cutle fish ink is used as a food ingredient as it is available in plenty to enhance flavor of food. It is also used as a food preservative to enhance shelf life because of its antimicrobial properties. Antimicrobial property of the cephalopod ink is well documented and there are several publications supporting this property (Takai *et al.* 1993., Sheu and Chou 190, Nirmale *et al.* 2002, Chacko and Patterson,2005, Vennila *et al.* 2011).

15.6 REFERENCES

1. Aristotle, The History of Animals, Book IV (ca. 350 BC). Translated by Arthur Leslie Peck and Edward Seymour Forster. Aristotle XII: Parts of Animals Movement of Animals, Progression of Animals (1937).

2. Bush, S.L.; Robison, B.H. Ink utilization by mesopelagic squid. Mar. Biol. 2007, 152, 485–494. [Google Scholar] [CrossRef].

3. Chacko, D.; Patterson, J. Effect of Pharaoh cuttlefish, Sepia pharaonis ink against bacterial pathogens. Indian J. Microbiol. 2005, 45, 227–230.

4. Charles D. Derby Cephalopod Ink: Production, Chemistry, Functions and Applications. A review. Mar. Drugs2014, 12(5), 2700-2730; doi:10.3390/md12052700.

5. Clancy, C.M.; Simon, J.D. Ultrastructural organization of eumelanin from Sepia officinalis measured by atomic force microscopy. Biochemistry 2001, 40, 13353–133

6. Hanlon R.T., Messenger J.B. Cephalopod Behaviour. Cambridge University Press; Cambridge, UK: 1996.

7. Hoyle W.E. Report of the Cephalopoda collected by H.M.S. Challenger during the years 1873–1876. Zoology. 1886;16:1–245.

8. Hu M.Y., Sucré E., Charmantier-Daures M., Charmantier G., Lucassen M., HimmerkusN.andMelzner F. Localization of ion-regulatory epithelia in embryos and hatchings of two cephalopods. Cell Tissue Res. 2010;339:571–583. doi: 10.1007/s00441-009-0921-8.

9. Laurie M. The organ of Verrill in Loligo. Quart. J. Microsc. Soc. Lond. 1888;29:97–100.

10. Matsuura, T.; Hino, M.; Akutagawa, S.; Shimoyama, Y.; Kobayashi, T.; Taya, Y.; Ueno, T. Optical and paramagnetic properties of size-controlled ink particles isolated from Sepia officinalis. Biosci. Biotechnol. Biochem. 2009, 73, 2790–2792.

11. Müller H. Bau der Cephaopoden. Z. Wiss. Zool. 1853;4:337–358.

12. Nair, J.R.; Pillai, D.; Joseph, S.M.; Gomathi, P.; Senan, P.V.; Sherief, P.M. Cephalopod research and bioactive substances. Indian J. Geo-Mar. Sci. 2011, 40, 13–27.

13. Nirmale, V.; Nayak, B.B.; Kannappan, S.; Basu, S. Antibacterial effect of the Indian squid, Loligoduvauceli (d'Orbigny) ink. J. Indian Fish. Assoc. 2002, 29, 65–69.

14. Palumbo, A. Melanogenesis in the ink gland of Sepia officinalis. Pigment. Cell Res. 2003, 16, 517–522

15. Prota G. Melanins, melanogenesis and melanocytes: Looking at their functional significance from a chemist's viewpoint. Pigment. Cell Res. 2000; 13:283–293.

16. Prota G., Ortonne J.P., Voulot C., Khatchadourian C., Mardi G., Palumbo A. Occurrence and properties of tyrosinase in the ejected ink of cephalopods. Comp. Biochem. Physiol. B. 1981;68:415–419.

17. Prota, G.; Ortonne, J.P.; Voulot, C.; Khatchadourian, C.; Mardi, G.; Palumbo, A. Occurrence and properties of tyrosinase in the ejected ink of cephalopods. Comp. Biochem. Physiol. B 1981, 68, 415–419.

18. Schmitt, C.B. Aristotle as a cuttlefish: The origin and development of a Renaissance image. Stud. Renaiss. 1965, 12, 60–72. [Google Scholar] (CrossRef)

19. Sheu, T.-Y.; Chou, C.C. Antimicrobial activity of squid ink. J. Chin. Agric. Chem. Soc. 1990, 28, 59–68.

20. Takai, M.; Yamazaki, K.; Kawai, Y.; Inoue, N.; Shinano, H. Effects of squid liver, skin, and ink on chemical characteristics of "ika-shiokara" during ripening process. Bull. Jap. Soc. Sci. Fish. 1993, 59, 1609–1615.

21. Vennila, R.; Rajesh, R.K.; Kanchana, S.; Arumugam, M.; Balasubramanian, T. Investigation of antimicrobial and plasma coagulation property of some molluscan ink extracts: Gastropods and cephalopods. Afr. J. Biochem. Res. 2011, 5, 14–21. 132

22. Verrill A.E. The cephalopods of the north-eastern coast of America. Part II. The smaller cephalopods, including the squids and the octopi, with other allied forms. Trans. Conn. Acad. Sci. 1880;5:259–446.

23. Von Boletzky S. Puffing smoke-rings underwater: The functional morphology of cephalopod ink ejectors. Vie Milieu. 1997;471:180–181.

24. Von Boletzky, S.,Juvenile behavior. In Cephalopod Life Cycles; Boyle, P.R., Ed.; Academic Press: London, UK, 1987; Volume II, pp. 45–84. (Google Scholar).

25. Voss G.L. Function and comparative morphology of the funnel organ in cephalopods; Proceedings of the XVI International Congress of Zoology; Washington, DC, USA. 20–27 August 1963; p. 74.

26. Wang F.R., Xie Z.G., Ye X.Q., Deng S.G., Hu Y.Q., Guo X., Chen S.G. Effectiveness of treatment of iron deficiency anemia in rats with squid ink melanin-Fe. Food Funct. 2014;5:123–128. doi: 10.1039/c3fo60383k.

27. Weiss F.E. On some oigopsidcuttle fishes. Quart. J. Microsc. Soc. Lond. 1888;29:75–96.

28. Young R.E., Mangold K.M. Cephalopod Pseudomorph Function. Tree of Life Web Project. [(accessed on 24 April 2014)]. Available online: http://tolweb.org/accessory/Cephalopod_Pseudomorph_Function?acc_id=1964

29. Zhong, J.P.; Wang, G.; Shang, J.H.; Pan, J.Q.; Li, K.; Huang, Y.; Liu, H.Z. Protective effects of squid ink extract towards hemopoietic injuries induced by cyclophosphamine. Mar. Drugs 2009, 7, 9-18.

CHAPTER **16**

AMBERGRIS

Ambergris (/ÈæmbYrariÐs/ or /Èæmb Yrarjs, Latin: *ambragrisea*, Old French: *ambregris*), *ambergrease* or *grey amber*, is a solid, waxy, flammable substance of a dull grey or blackish colour, produced in the digestive system of sperm whales (Encyclopedia Britannica 2013). In Eastern cultures ambergris is used for medicines and potions and as a spice; in the West it was used to stabilize the scent of fine perfumes. Ambergris floats and washes ashore most frequently on the coasts of China, Japan, Africa, and the Americas and on tropical islands such as the Bahamas. Because it was picked up as drift along the shores of the North Sea, ambergris was likened to the amber of the same region, and its name is derived from the French words for "gray amber." Fresh ambergris is black and soft and has a disagreeable odour. When exposed to sun, air, and seawater, however, it hardens and fades to a light gray or yellow, developing a subtle and pleasant fragrance in the process" (Encyclopædia Britannica, 2013).

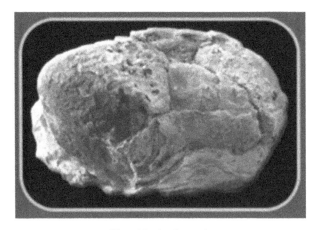

Fig. 46: Ambergris

16.1 OCCURRENCE

According to a recent report a couple in the UK who got 1.57 kg ambergris have entered into negotiations for selling at a price ofUS$70,000 (AU$100,000 (BecCrew14 April 2016).According to National Geographic, it's illegal to use ambergris to produce perfumes in the US, due to sperm whales being endangered, but it's still a big thing in France - Lanvin and Chanel famously use the substance to make their fragrances last longer on the skin. Ambergris has a scent all its own—derived from its chemical component ambrein—that it imparts to popular perfumes such as Chanel No. 5(Cynthia Graber on April 26, 2007). In India an ambergris weighing 15 kg was netted by fishermen of the Chinnoor coastal village, in Tamil Nadu State, in the mid-sea. According to customs officials this piece of ambergris is expected to cost1.5-2 millions in Indian market and Rs. 7.5 millions in the international markets.

Fig 47: Picture of Ambergris
Photo: The Hindu (The National Newspaper, India dated February 28, 2013.

16.2 THE ORIGIN OF AMBERGRIS

Early and modern theories of the origin of ambergris are described in an excellent review by Robert Clarke (2006). Ambergris occurs in both male and female sperm whales, and also in the pygmy sperm whale. It occurs in about one in 100 sperm whales. Its origin in literature dates back to 9th century to an Arab traveler who cites its trade in islands of Indian Ocean(KERR,1718).Chevalier (1700) recorded in his writing that the King of Tudor in the Mollucas gifted a 85 kg boulder of precious ambergris to Dutch East India Company. The largest find ever recorded weighed 455kg and sold for £23,000 in 1914 (Clare, 2006).

Ambergris occurs in the rectum of the whale but neither causes nor betrays disease. The rectum is not damaged by squid beaks. Indigestible material, that is, squid beaks and pens and the cuticles of parasitic nematode worms, are regularly vomited by sperm whales and the intestine and rectum can only deal with liquid faeces. When, as sometimes happens, some indigestible material leaks into the intestine and, by at least partly blocking the flow of the faeces. The tangled mass is pushed into the rectum where there is reason to believe that the water absorbing capacity of the rectum is increased. In this way the faecal matter is precipitated on the indigestible material to form a smooth concretion and the faeces can pass again. Then more faecal material arrives and the process is repeated. In this way the flow of liquid faeces is maintained, although at the expense of accretionary growth in size of the coprolith which becomes ambergris. Response processes in the whale are constructional. The biochemical processes which transform the coprolith into ambergris are consequential upon its incubation over a long period in the peculiar environment of the rectum teeming with bacteria. Eventually the rectum stretches until it breaks, causing the whale's death and the ambergris is released into the sea (Clarke, 2006). It is also called "Preternaturally hardened whale dung"and *also "Floating Gold"* (Christopher Kemp, 2012). Although several theories have been propagated about the origin of ambergris in sperm whale the most widely accepted theory is that of Lambertsen and Kohn (1987) that "ambergris in the sperm whale forms as a pathological concretion of faecal material."

16.3 MAIN CONSTITUENTS OF AMBERGRIS

Ambergris contains alkaloids, acids and a main compound similar to cholesterol called ambrein which on oxidation produces two compounds called ambroxan and ambrinol. Of the three compounds only ambrein has the typical scent.

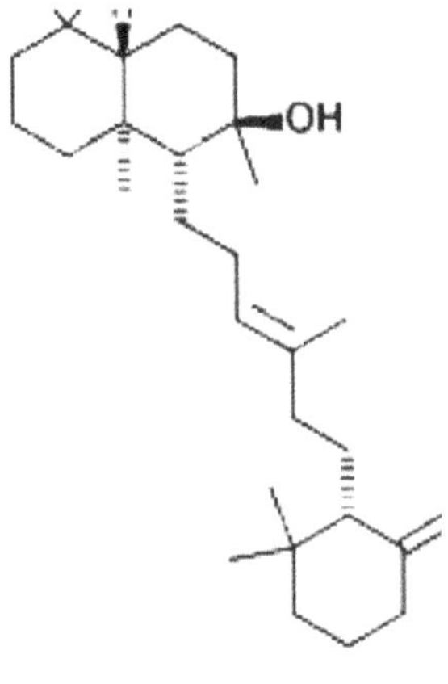

Fig. 48: Ambrein

According to Lederer (1949) the principal constituents are ambrene ($C_{10}H_{52}O$) 25-45%, epi- coprosteroil (free and esterified) 30-40%, coprosterol, 1-5%, cholesterol 0-1%, ketones (more than 50% coprostane-3-one) 6-85, free acid 5%, esterified acids 5-8% a hydrocarbon pristine ($C_{18}H_{38}$) 2-4%.

16.3.1 Inorganic constituents

Takeo Ishiguro *et al.* (1952) showed that main inorganic metallic residues are: Ga, Fe*, K*, Mg, Na, (P), Si, Sr*, Zn; Group II - Al, Nb*, Cr*, Cu, La*, Mn, Ni*, Ti*, W* (where* denotes presence in minute amounts). The quantitative analysis of the ash gave the following results. CaO 6.21%, MgO 9.88%, P_2O_5 4.65%, SiO_2 6.02%.

16.4 USES OF AMBERGRIS

Ambergris, is a metabolite of the sperm whale. It is extensively used in perfume industry as a fixative. It considered one of the most remarkable animal perfumes. The autoor photo-oxidative decomposition of (+)-ambrein, the major constituent of ambergris, gives the five important odoriferous products that play a key role in the characteristic odor of ambergris. However, the unique olfactive and fixative properties are related principally to (-)-ambrox, one of the decomposing products of (+)-ambrein and the commercially most important synthetic equivalent of ambergris, which has been successfully synthesized from various terpenods (Wang *et al.*, 2001).

In order to understand how it is used in perfume industry I went to U. K and visited in 1984 in the house of Dr. Fred Wells world renowned Perfume Chemist and held discussions with him about the mode of application of ambergris in perfume industry. He told me that a tincture of 3-5% ambergris is prepared in absolute alcohol and kept stored in darkness for more than six years. During this period, it undergoes several chemical changes. Afterwards it is taken out and used as a fixative. Normally a perfume contains several chemical perfumes. The role of a fixative is to blend them all together and give entirely a new characteristic smell and the formula is a trade secret. According Dr. Wells ambergris is by far the best fixative ever known to man and was in use from time immemorial by perfume industry.

16.5 REFERENCES

1. "Ambergris". Britannica. Retrieved, 31January 2013. (https://en.wikipedia.org/wiki/Ambergris) (Ambergris Vol. 1, Chap. Iv, P. 47).

2. A.V.Ragunathan. Fishermen net ambergris in mid-sea. Cuddalore, February 28, 2013 04:28 IST). DOI: http://dx.doi.org/10.5597/lajam00087.

3. CHEVALIER, N. (1700) Description de la pièce D'Ambre Gris que la chambred'amsterdam a recenë des Indes.

4. Christopher Kemp. 2012 Floating Gold: A Natural (and Unnatural) History of Ambergris.

5. Clarke, Robert (2006). "The origin of ambergris". Latin American Journal of Aquatic Mammals.5 (1): 7–21. Doi:10.5597/lajam00087.

6. Cynthia Graber on April 26, 2007. Strange by True: Whale Waste is Extremely Valuable. Scientific American. https://www.scientificamerican.com/article/strange-but-true-whale-waste-is-valuable/

7. Floating Gold: A Natural (and Unnatural) History of Ambergris. University of Chicago Press, 2012 187 pages.

8. Inorganic Constituents of Ambergris yakugakuzasshi, Volume 72 (1952) Issue 11 Pages 1439-1443., Doi: Https://Doi.Org/10.1248/Yakushi1947.72.11_1439.

9. Johnna Rizzo, 2012.What's Ambergris? Behind the $60k Whale-Waste Find. National Geographic News. https://news.nationalgeographic.com/news/2012/08/120830-ambergris-charlie-naysmith-whale-vomit-science/. Last accessed: Jan 2018.

10. Kerr, R. (Edit.) (1718) A General History and Collection of Voyages and Travels. Republished 1811-24, Edinburgh.

11. Lambertsen, R.H. And Kohn, B.A. (1987). Unusual Multisystematic Pathology in A Sperm Whale Bull.Journal of Wildlife Diseases 23: 510-514.

12. Lederer, E. (1949) Chemistry and Biochemistry of some Mammalian Secretions and Excretions. Journal of the Chemical Society, Part 3: 2115-2125.

13. Orientales, pesant 182 livres; avec un petit traité de son Origine et de savertu. Amsterdam, Chez l'Anteur.

14. Perraudin, Frances (13th April 2016). Lancashire couple hoping to cash in on 'what vomit' windfall. https://www.theguardian.com/uk-news/2016/apr/13/lancashire-couple-whale-vomit-ambergris-middleton-sands. Last accessed: Jan 2018.

15. Takeo Ishiguro, Naofumi Koga And Yoshimasa Matsuo Studies on The Constituents of Ambergris.

16. Wang Wenjun, chensha and Dai Qianhuan. Composition of ambergris and the development of synthesis of ambrox[J]. Chin. J. Org. Chem., 2001, 21(3): 167-172.

CHAPTER 17

AIR BLADDER OF FISH/ FISH MAWS

According to Medical Dictionary "air bladder is a gas-filled sac that is present in most fish and functions as a hydrostatic organ. It is located beneath the vertebral column primarily in the anterior abdomen and is connected with the esophagus in some species (for example, goldfish). Oxygen is transferred from a rich venous sinus into the swim bladder to increase buoyancy. Synonym(s): swim bladder".

For more than a millennium, fish maw has been a favorite dish at fancy Chinese banquets, celebrated in large part for its supposedly medicinal properties. Fish maw existed in the Chinese dishes since 206 BCE (Han Dynasty). According to records the Chinese chefs prepared fish maw dishes for the emperors in grand banquets at celebrations for the family and when important dignitaries came to visit. In modern days fish maw dishes are in China's state dinners for visiting leaders. This classic dish was grandly presented to the state dinner 2014 CICA Shanghai Summit. (Conference on Interaction and Confidence-Building Measures in Asia). Now-a-days fish maw dishes have become ever popular on Chinese dining tables especially for occasions like celebration for birthdays, weddings, Chinese New Year etc. The Chinese people believe that drinking fish maw soup and eating fish maw will improve their skin beauty and, for pregnant women, the skin of their babies. Furthermore, fish maw does not contain cholesterol. It is therefore being a premium health enhancing ingredient suitable for long time consumption.

The bigger the size of the bladder, the steeper the price. Those sliced from smaller species on the croaker family tree were commonly used in cooking, while the much larger specimens that came from the giant yellow croakers cost many times more and were consumed typically for specific medical benefits.

This over fishing decimated the croaker population. In the late 1930s, boats hauled up 50,000 tonnes of giant yellow croaker a year; three decades later, it was as little as 10 tonnes. Another effect was to sharply reduce the average size of the croakers caught. As a result, the price of large bladders soared. While in the 1930s, they sold for a few US dollars, a large bladder cost as much as $1,000 ($6,500 in 2015 dollars) in 1969. By 2000, they were going for $21,800. In real terms, that means the price multiplied nearly five-fold in three decades.

"A fisherman can make $3,000 to $4,000 for a swim bladder—it's a good living to catch just one,".

17.1 FISH MAW

The fish maw is the dried swim bladders that come from a large fish like a croaker or sturgeon. In South east Asian countries like China, Hong Kong, Japan, Singapore etc. it is considered to be one of the top delicacy. In appearance, dried fish maw is light, white in color, and has a spongy texture. Dried fish maw is tasteless which makes it a good complementary addition to many dishes since it can absorb the flavors of other ingredients when it is cooked with other food ingredients.

17.1.1 How to use fish maws for making food preparations

Dried fish maw must first be rinsed and then soaked in a large bowl of water to soften the texture before use. There should be enough water to allow room for the fish maw to expand. The period of soaking typically ranges from 6 to 12 hours or until it becomes softened. Afterward, squeeze out the water from the fish maw to rid it of any smell. Next the fish maw can be cut into smaller pieces to make it easier for consumption. Now it's ready to be cooked or added to the soups or dishes of your choices. After being cooked the fish maw will have a soft and slippery texture.

17.2 HEALTH BENEFITS

Fish maw is a good source of collagen, proteins, and nutrients. Collagen provides a wide variety of skin benefits, such as helping to improve your skin tone, and tissue health. Furthermore, many Chinese people considered the fish maw as a traditional delicacy that represents fortune and health. Therefore, it is popularly served during special occasion such as Chinese New Year, birthdays, and weddings (eatconnection.com). Some of the health benefits are listed below (http://oceanwild.co.nz/the-story-of-fish-maw).

- Fish maw contains rich proteins and nutrients.

- It nourishes 'yin', replenishes kidney and boosts stamina.

- It is effective in improving weak lung and kidney, anemia etc.

Fish maw is a good source of collagen, which is vital in healthy skin beauty in all age group of either gender, for pregnant women, the skin of their babies. Collagen is capable of preserving fine complexion and invigorating blood circulation. This is significant in maintaining and prolonging your youthful, radiant skin. *Health definitely is wealth to everyone. And fish maw dishes are believed to be flavor, fortune, healthy.*

17.3 FISH MAW FOR CLARIFICATION OF WINE

The process of clarification of wine is called fining. Fining is the process of purifying wine of unwanted tannins, sediment, and colors. "Fining, the process used by a large portion of winemakers to purify and stabilize the wine, gives it clarity of color, removes sediment and suspended solids, and strips away any unwanted tannins, odors, or colors. It's one of the most influential steps on the outcome of the finished product" (Thomas Keller). Some of the materials widely used include casein (milk protein), egg whites, gelatin (taken from pig or cow skin and connective tissue), chitosan (crustacean exoskeletons), kieselsol (colloidal silicic acid) and is in glass (fish swim bladder). Fining agents such as isinglass, chitosan, and casein are almost exclusively used for white wines, and egg whites are used for red wines.

17.4 HOW TO PREPARE ISINGLASS FOR REFINING WINE?

Dried fish mass is converted to very thin flakes and suspended in ice water for several hours. Addition of small amounts of citric acid is good. The swollen liquid is used for fining. Addition of 1 cc of this isinglass is sufficient to clarify 1 litre of wine. The wine become crystal clear on keeping the wine with added isinglass emulsion after 3-4 hrs or preferably overnight.

17.5 REFERENCES

1. https://eatconnection.com/what-is-fish-maw/

2. http://oceanwild.co.nz/the-story-of-fish-maw/

3. https://munchies.vice.com/en_us/article/ypxjzg/theres-blood-and-bladders-in-your-wine

CHAPTER **18**

STAR-FISH

S tarfish are marine invertebrates. It is considered as the oldest species lived on earth, more than 450 million years ago. Starfish or sea stars are star-shaped echinoderms belonging to the class *Asteroidea*. About 1,500 species of starfish occur on the seabed in all the world's oceans, from the tropics to frigid polar waters. They are found from the intertidal zone down to abyssal depths, 6,000 m (20,000 ft) below the surface (Wikipedia). Starfish have a central disc and five arms, though some species have a larger number of arms. The aboral or upper surface may be smooth, granular or spiny, and is covered with overlapping plates. Many species are brightly coloured in various shades of red or orange, while others are blue, grey or brown.

Fig. 49: Star fish in natural habitat
(Under water photo courtesy B.M. Noushad)

18.1 STAR FISH AS FOOD

They are seldom consumed as a food as most of them often contain saponins and tetradotoxins which are extremely poisonous. But there are some reports showing that people of some region consume some species which may not be toxic in that region of ocean. George BerhardRumpf found few starfish being used for food in the Indonesian archipelago, other than as bait in fish traps, but on the island of "Huamobel" the people cut them up, squeeze out the "black blood" and cook them with sour tamarind leaves; after resting the pieces for a day or two, they remove the outer skin and cook them in coconut milk and consumed. Starfish are sometimes eaten in China (Rumphius 1999),Japan (The China Guide 2011; Amakusa 2013; Pouch 2013) and in Micronesia (Johannes 1981).

18.2 STARFISH AS A SOURCE OF COLLAGEN

Collagen from fish is now considered a safe material compared to gelatin from cattle. Collagen is regarded as one of the most useful biomaterials. The excellent biocompatibility and safety due to its biological characteristics, such as biodegradability and weak antigenecity, made collagen the primary resource in medical applications. "The main applications of collagen as drug delivery systems are collagen shields in ophthalmology, sponges for burns/wounds, mini-pellets and tablets for protein delivery, gel formulation in combination with liposomes for sustained drug delivery, as controlling material for transdermal delivery, and nanoparticles for gene delivery and basic matrices for cell culture systems. It was also used for tissue engineering including skin replacement, bone substitutes, and artificial blood vessels and valves" (Chi H. Lee *et al.*, 2001).

Fish collagen has been an area of intense research in recent times. Studies on extraction and properties of collagen from fish. Star fish (Asteriasamuensis) by Kj-jeong Lee *et al.* (2001), Baltic cod (Gadusmorhua) by Maria Sadowska *et al.* (2003), from bones and scales of blackdrum (Pongoniacromis) by Mashiro *et al.* (2004), muscle of skate (Rajakenojei) by Shoshi *et al.* (2002) and from purple sea urchin (Anthocidanscrassispina) by Takashi Nagi and NobutakSuzyki (2000) are some of the notable investigations in this field.

Maya Raman, 2005 have shown that the epidermal connective tissue of India squid (Lologoduvauceli, Orbigny) have high collagen content (17.8%). They have identified by SDS-PAGE electrophoresis that this collagen consists of alpha, beta and gama sub-chains. The same author earlier observed that collagen content of shark (24.85%) was higher that freshwater fish Rohu (17.3%) of their total protein wet weight basis.

Some of the other important biomedical applications of collagen include to make bandages for burn/wound cover dressings, bone filling materials, antithrombogenic surfaces, biodegradable sutures and immobilization of therapeutic enzymes. Recently collagen is used also as a carrier for drug delivery systems. Ka-jeong Lee *et al.* (2009) have shown that pepsin solubilized collagen (PSC) isolated from the tissue of a starfish (Asteriasamurensis) consisted of (á1)2á2 heterotrimer, which is similar to calf skin type I collagen. However, the PSC was denatured at 24.7° C, which is about 12° C lower than mammalian collagen. Immunoblotting assay using polyclonal anti-type I collagen antibody revealed that the starfish collagen contained similar affinity motifs. In XTT assay, PSC suspension had cell growth activity and showed no cytotoxicity.

18.3 REFERENCES

1. Johannes R.E., Words of the Lagoon: Fishing and Marine Lore in Palau District of Micronesia (Berkeley: Univ. of California Press, 1981).

2. Lee, C.H.; Singla, A.& Lee, Y. (2001). Biomedical applications of collagen. International Journal of Pharmaceutics 221(June 2001) 1–22.

3. Lee, K.-j., Park, H. Y., Kim, Y. K., Park, J. I., & Yoon, H. D. (2009). Biochemical Characterization of Collagen from the Starfish (Asteriasamurensis). J. Korean Soc. Appl. Biol. Chem, 52(3), 221-226.

4. Maya Raman, 2005. Ph.D. Thesis, Cochin University of Science and Technology, Kochi, India.

5. Mizuta, S., Hwang, J., & Yoshinaka, R. (2002). Molecular species of collagen from wing muscle of skate (Raja kenojei). Food Chemistry, 76(1), 53-58.

7. Nagai T., & Suzuki, N. (2000). Preparation and characterization of sever fish bone collagens. Journal of Food Biochemistry 24(5) 427-436

7. Ogawa, M., Portier, R. J., Moody, M. W., Bell, J., Schexnayder, M. A., & Losso, J. N. (2004, December). Biochemical properties of bone and scale collagens isolated from the subtropical fish black drum (Pogonia cromis) and sheepshead seabream (Archosargus probatocephalus). Food Chemistry, 88(4), 495-501.

8. Rigby, B. J. (1968, July). Amino-acid Composition and Thermal Stability of the Skin Collagen of the Antarctic Ice-fish. Nature, 219, 166-167.

9. Rumphius, Georgious Everhardus; Beekman, E.M. (trans.) (1999) [1705]. The Ambonese Curiosity Cabinet (original title: Amboinsche Rariteitkamer). Yale University Press.

10. Sadowska, M., Kolodziejska, I., & Niecjkowska, C. (2003, May). Isolation of collagen from the skins of Baltic cod (Gadusmorhua). Chemistry, 81(2), 257-262.

CHAPTER **19**

FISH COLLAGEN

ollagen is the most abundant protein in our body and is a vital component for building several membranes. But our bodies produce less and less of it with aging. Since most of us do not get enough collagen through our diets, supplementation through diet with a quality collagen is essential to keeping our bodies strong and healthy. There are several marine fish with high collagen content. Some of them are in proteins. Collagen from fish has several good health benefits compared to collagen from plants. These include helping to decrease wrinkles and scars, stabilizing blood sugar, reducing inflammation and rebuilding of cartilages. It is also source of high quality protein that is easily absorbed and effective to reverse aging from the inside out.

19.1 DIFFERENCES BETWEEN COLLAGEN OF MARINE ANIMALS AND LAND ANIMALS

There are more than 40 types of collagens identified existing in mammalian bodies. Among the collagens identified the richest one present in human body is the type 1. Type 1 collagen is the one seen in skin, bones, teeth, tendons ligaments and facia. Marine collagen is mainly extracted from fish skin, intestines and other wastes. Fish collagen is rich in essential amino acids. Fish collagen contains 8 out of 9 essential amino acids. Glycine and proline are present in very high quantities. Glycine is biologically very important. It is vital for the performance of several biological functions in human body like building of bones, and muscles. Glycine plays an important role in giving messages to nervous system to prevent formation of ulcers in stomach. It is also needed to improve glucose tolerance to diabetic patients. Proline is a scavenger of free radicals and hence an antioxidant and also prevent cell damages.

19.1.1 Fish collagen

In the fish processing industries 30 to 40 % are wastes. Several by-products are now prepared today. The amount of wastes varies from species to species (Chalamaiah *et al.* 2012). This include chitin, chitosan, chondroitin sulphate, gelatin, collagen, glucose amine etc. Among the fish wastes skin and intestines are rich source of collagen. Several research scientists hasve worked out the methods of extraction and purification of fish collagen and biological applications also.

Health benefits of marine collagen

Maya Raman (2005) has worked out amino acid composition of collagen fractions of squid skin, tentacle and mantle and shown that almost all amino acids are present in the tentacle and mantle (see table below).

Table 31: Amino acid profile of collagen fractions of squid (*Loligo duvacelli*) skin, tentacle and mantle

Amino acids	Epidermal layer	Tentacle	Mantle
Aspartic acid	10.18	23.12	42.08
Threonine	17.78	10.05	16.61
Serine	–	9.61	15.32
Glutamic acid	–	46.72	95.40
Proline	–	3.92	6.52
Glycine	–	19.15	18.15
Alanine	–	17.74	29.99
Cystine	20.08	1.83	–
Valine	–	11.38	19.84
Methionine	18.04	18.72	34.20
Isoleucine	34.85	57.07	173.96
Leucine	10.19	19.30	24.51
Tyrosine	9.52	12.60	20.73
Phenylalanine	–	1.34	0.97
Histidine	5.78	23.22	36.41
Lysine	42.05	41.99	81.34
Arginine	0.42	0.94	272.69

Bone loss is an important sign of old age. This leads to development of osteoporosis which is now a debilitating disorder, a health hazard to millions of people around the world. Women are more prone to osteoporosis to than man due to dietary habits, poverty, social customs and type of work they do. Osteoporosis is a condition that causes weakening of the bones in your body. It is also called "brittle bone disease,". Osteoporosis increases your chance of sustaining a broken bone. In our body new bones are made and old bones are taken away. This is a continuous process. Osteoporosis develops when the pace of new bone formation cannot match with the loss of bone. Calcium and phosphorus are two very important minerals needed for the bio-synthesis of bones and rebuilding of damaged cartilages. But absorption of these vital minerals in human body is very slow and this results in deficiencies. This eventually leads to bone loss in old age. Collagen peptides can facilitate the absorption of both calcium, phosphorus and other vital minerals. Studies also have shown that marine collagen has excellent properties to stimulate collagen synthesis in bones and promote bone building cells called osteoblasts. Osteoblasts are the cells that build the organic matrix of bones that is composed of 90% collagen. There are three types of cells in your body that bring about changes in our bones: osteoblasts, osteocytes, and osteoclasts. *"Osteoblasts are bone forming cells. Of the three types of bone cells, they are the ones that produce the matrix that makes up bone. The matrix, or organic material, includes molecules such as collagen protein fibers, which give bone its flexibility, and calcium (Ca2+) and phosphate (PO4-) ions, which give bone its rigidity. Osteoblasts make and package the matrix molecules for release into the extracellular environment. Once released, the molecules in the matrix react with each other to form a rigid yet flexible bone tissue called* **osteoid** *that eventually hardens to form bone"* (Catherine Konopka, 2017).

19.2 SOME OF THE USES OF COLLAGEN

19.2.1 Anti-aging properties of collagen

About 70% of human skin is made up of Type 1 collagen as the process of aging sets in skin collagen starts breaking. This is a continuous process. Another process taking place during aging is the development drying of the skin and getting saggy. Wrinkles start appearing in the body particularly in the face. Supplementation of skin with can greatly prevent formation of wrinkles. This is one reason most of the anti-aging face and body lotion contain collagen as a principal component. These days collagen is in high demand from cosmetics and medical companies.

19.2.2 Collagen and diabetics

It is now recognized that low levels of glycine in blood result impaired glucose tolerance and leads to insulin resistance leading to the onset of type II diabetics (Richard Yan-D 2017). Therefore, it is now suggested that supplementation of diets rich in glycine containing protein can help to regulate and balance blood glucose level. But this needs more confirmation from further research in this field.

19.2.3 Wound healing properties of collagen

Collagen id found to hasten wound healing and regeneration of tissues. Collagen peptides are now in use as bandages for wound healing and as dressing burn injuries, help protein synthesis and extracellular matrix for tissue synthesis (Brett, 146

2008). Fresh sterilized fish skin is now used as dressing to hasten injuries due to burn and promote skin formation and reduce healing time. Tilapia skin is now extensively used in many countries.

19.2.4 Collagen based medical products

One of the great advantages of collagen is that it can be purified to a great extent., It is biocompatible and can be converted to a variety of shapes and sizes, and acts as a temporary scaffold for tissue regeneration. Modern collagen processing allows to make biocompatible medical products in many different forms I like sheets, putties, injectables, gels, three dimensional shapes, coatings, fibers, powders and soluble sutures.

19.2.5 Collagen blended products

Different collagen configurations have the potential to be used in a variety of clinical applications like bone void fillers, cartilage repair devices, wound dressings, haemostatic agents, nerve regeneration conduits and tissue augmentation supports.

One of the developments now popular is collagen blended medical products. Collagen can be blended with several other chemicals to enhance their biodegradability. Some of them available in the market and widely used are given below.

- Beta-tricalcium phosphate
- Hydroxyapatite
- Anorganic bone
- Calcium sulfate
- Bioactive glass
- Synthetic polymers
- Polysaccharides
- Hylauronic acid
- Alginates

Fish collagen and collagen peptides find application as antimicrobial agents instead of antibiotics and also as an anti-inflammatory agent.

19.3 REFERENCES

1. Brett, DW. (2008). A Review of Collagen and Collagen-based Wound Dressings.". Wounds, 20(12), 347-356.

2. Chalamaiah, M., Hemalatha, R., & Jyothirmayi, T. (2012). Fish protein hydrolysates : proximate composition, amino acid composition, antioxidant activities and applications: A review. Food Chemistry, 135(4), 3020-3038.

3. Cui-Feng, Z., Guan-Zhi, L., Hong-Bin, P., Fan, Z., Yun, C., & Yong, L. (2010). Treatment with marine collagen peptides modulates glucose and lipid metabolism in Chinese patients with type 2 diabetes mellitus. Applied Physiology, Nutrition, and Metabolism, 35(6), 797-804.

4. Konopka, C. (2017, Jan 28). Osteoblast: Definition, Function & Differentiation. Récupéré sur https://study.com/academy/lesson/osteoblast-definition-function-differentiation.html

5. Raman, M. (2005). Effect of thermal modification on the Rheological Characteristics and protein quality of three species of fish of varying collagen content. PhD Thesis. Cochin University of Science and Technology.

6. Maya Raman, 2005. effect of thermal modifications on the theological characteristics and protein quality of three species of fish of varying collagen content. Cochin University of Science and Technology, Cochin, Kerala, India.

7. Yamada, S., Nagaoka, H., Terajim, M., Tsuda, N., Hayashi, Y., & Yamauchi, M. (2013). Effect of fish collagen peptides on collagen post translational

modifications and mineralization in an osteoblastic cell culture system. Dental Materials Journal, 32(1), **88-95.**

8. Yan-Do, R., & MacDonald, P. E. (2017, May). Impaired "glycine"-mia in type 2 diabetes and potential mechanisms contributing to glucose homeostasis. Endocrinology, 158(5), 1064- 1073.

9. Zhu, C.F., Li, G.Z., Peng, H.B., Zhang, F, Chen, Y., Li, Y. (2010). Treatment with marine collagen peptides modulates glucose and lipid metabolism in Chinese patients with type 2 diabetes mellitus. Appl Physiol Nutr Metab. 35(6), 797-804.

INDEX